"Lushes sober up now and then and go about scowling at the world, but I never sober up. That's because I don't depend on alcohol, or any chemical substance, to get me lubricated. Life is one long high for me, and, in particular, writing these essays is enough to elevate me even in moments of difficulty.

So let's get on with it."

—Isaac Asimov

Provocative, enthralling, and consistently entertaining, in FAR AS THE HUMAN EYE CAN SEE, Isaac Asimov once again explores the fascinating territory he knows and loves best, willing to discuss anything and everything in the vast realm of science. A remarkable collection of reality's wonders by the author of the acclaimed FOUNDATION saga.

# THE PEOPLE BEHIND THE HEADLINES
## FROM ZEBRA BOOKS!

PAT NIXON: THE UNTOLD STORY         (2300, $4.50)
by Julie Nixon Eisenhower

The phenomenal *New York Times* bestseller about the very private woman who was thrust into the international limelight during the most turbulent era in modern American history. A fascinating and touching portrait of a very special First Lady.

STOCKMAN: THE MAN, THE MYTH,
THE FUTURE         (2005, $4.50)
by Owen Ullmann

Brilliant, outspoken, and ambitious, former Management and Budget Director David Stockman was the youngest man to sit at the Cabinet in more than 160 years, becoming the best known member of the Reagan Administration next to the President himself. Here is the first complete, full-scale, no-holds-barred story of Ronald Reagan's most colorful and controversial advisor.

IACOCCA         (1700, $3.95)
by David Abodaher

He took a dying Chrysler Corporation and turned it around through sheer will power and determination, becoming a modern-day folk hero in the process. The remarkable and inspiring true story of a legend in his own time: Lee Iacocca.

STRANGER IN TWO WORLDS         (2112, $4.50)
by Jean Harris

For the first time, the woman convicted in the shooting death of Scarsdale Diet doctor Herman Tarnower tells her own story. Here is the powerful and compelling *New York Times* bestseller that tells the whole truth about the tragic love affair and its shocking aftermath.

*Available wherever paperbacks are sold, or order direct from the Publisher. Send cover price plus 50¢ per copy for mailing and handling to Zebra Books, Dept. 107 , 475 Park Avenue South, New York, N.Y. 10016. Residents of New York, New Jersey and Pennsylvania must include sales tax. DO NOT SEND CASH.*

# ISAAC ASIMOV

## FAR AS HUMAN EYE COULD SEE

**PINNACLE BOOKS**
**WINDSOR PUBLISHING CORP.**

The essays in this volume are reprinted from the *Magazine of Fantasy and Science Fiction,* having appeared in the indicated issues:

Far as Human Eye Could See (November 1984)
Made, Not Found (December 1984)
Far, Far Below (January 1985)
Salt and Battery (February 1985)
Current Affairs (March 1985)
Forcing the Lines (April 1985)
Arise, Fair Sun! (May 1985)
The Rule of Numerous Small (June 1985)
Poison in the Negative (July 1985)
Tracing the Traces (August 1985)
The Goblin Element (September 1985)
A Little Leaven (October 1985)
The Biochemical Knife-Blade (November 1985)
The Discovery of the Void (December 1985)
Chemistry of the Void (January 1986)
Time is Out of Joint (February 1986)
Superstar (March 1986)

PINNACLE BOOKS

are published by

Windsor Publishing Corp.
475 Park Avenue South
New York, NY 10016

First Pinnacle Books printing: April, 1988

Printed in the United States of America

SCIENCE ESSAY COLLECTIONS BY ISAAC ASIMOV

## FROM THE MAGAZINE OF FANTASY AND SCIENCE FICTION

FACT AND FANCY

VIEW FROM A HEIGHT

ADDING A DIMENSION

OF TIME AND SPACE AND OTHER THINGS

FROM EARTH TO HEAVEN

SCIENCE, NUMBERS AND I

THE SOLAR SYSTEM AND BACK

THE STARS IN THEIR COURSES

THE LEFT HAND OF THE ELECTRON

THE TRAGEDY OF THE MOON

ASIMOV ON ASTRONOMY

ASIMOV ON CHEMISTRY

OF MATTERS GREAT AND SMALL

ASIMOV ON PHYSICS

THE PLANET THAT WASN'T

ASIMOV ON NUMBERS

QUASAR, QUASAR, BURNING BRIGHT

THE ROAD TO INFINITY

THE SUN SHINES BRIGHT

COUNTING THE EONS

X STANDS FOR UNKNOWN

THE SUBATOMIC MONSTER

FAR AS HUMAN EYE COULD SEE

## FROM OTHER SOURCES

ONLY A TRILLION

IS ANYONE THERE?

TODAY AND TOMORROW AND—

PLEASE EXPLAIN!

SCIENCE PAST—SCIENCE FUTURE

THE BEGINNING AND THE END

LIFE AND TIME

CHANGE!

THE ROVING MIND

THE DANGERS OF INTELLIGENCE

To SAM VAUGHAN who, when told once,

> *"Issac has just handed in an autobiographical manuscript twice the permitted length for a volume,"*

nobly replied, "Do it in two volumes!"

# CONTENTS

# INTRODUCTION

I have, as of now, written 329 science essays for the *Magazine of Fantasy and Science Fiction (F & SF)*, one each of 329 monthly issues, without missing a single one, for the last $27^5/_{12}$ years. All but six of these essays (six early ones, written before I had hit my stride) are to be found in one collection or another—some in more than one. In this collection we have essays #313 to #329 inclusive.

Writing that many is not an easy job, you may be sure, even for someone who loves writing as much as I do, and who finds it as easy as I do.

For one thing, what if you start repeating yourself? It is impossible to avoid this altogether. After all, each essay has to be as self-contained as possible, for it appears in an individual issue of a magazine and that issue may be the only one a particular reader will see. Therefore, I often have to explain something I have already explained in another essay. Sometimes, if the matter is tantential, I can get away with a parenthesis or a footnote directing the reeader to another essay in the collection, or even in another collection. If the matter is of the essence, I just explain again.

But what if, without realizing it, I were to rewrite an entire essay that I had already written before? I actually did this during the period I was writing the seventeen essays included in this collection. You'll find the frightening story (frightening to me, anyway) in the introductory paragraphs of chapter 6.

This time I caught it before it was too late, but the time will come inevitably (if I live long enough, and if my brain deteriorates sufficiently in consequence) when I will repeat an earlier essay and *not* catch it. And if our Noble Editor doesn't catch it (and why should he?) it will be published. And then about a thousand Gentle Readers will write to me to tell me what I have done, and a few of the less Gentle ones will surely mutter something about senile dementia, or, as it is now called, "Alzheimer's disease." (Poor Dr. Alzheimer, what a way to achieve immortality!)

Even if we put that to one side, what about the matter of achieving a decent balance in these essays?

When I was first asked to do these essays by *F & SF*, when the world and I were young—anyway, younger—I was told I would have a completely free hand in choosing subjects, as long as I thought that the subject would be of interest to the readers of the magazine. Naturally, the subject I was expected to choose most frequently was one that was scientific in nature, since the terms of the agreement defined the product as a "science essay."

That didn't bother me at all. I am, needless to say, endlessly interested in science and so, I firmly insist, are science fiction readers. Still, I have taken advantage of my *carte blanche* to write, occasionally, essays that are primarily concerned with history, or with sociology, or with my own opinions on this or that. I ahve written some that are nothing more or less than autobiographical.

This doesn't happen often, to be sure, but the magazine has stuck to its promise, loyally. Never once— never *once*—has an essay been sent back. Nor have I ever been asked to modify a single sentence.

And yet we may omit these off-the-wall essays, for the fact of the matter is that some 95 percent of these essays are about one branch or another of science.

Now the question is: Do I balance the various branches of science? Do I sit down at the typewriter, check some intricate mathematical formula, and say, "Aha, it is time to do an article on biophysics—or anthropology—or astrochemistry"?

I can't do that. It would make the whole process too difficult and too mechanical.

The technique I follow, when the time of the month rolls around, is to sit down at the typewriter and ask myself what I *feel* like writing about. Sometimes I know right off and sometimes I have to think a bit, but wherever the feeling leads me that's where I go.

The result is that the balance is skewed. Some branches of science interest me more than others, and those that interest me are represented perhaps more than they deserve.

I have never really made a statistical analysis of my 2¼ dozen dozen essays, but I strongly suspect that there are more essays on astronomical subjects than on subjects representing any other branch of science. After all, astronomy is my favorite science. I have

11

never taken a single astronomy course in college or graduate school, but as I have been a science fiction fan for better than half a century, astronomy *must* be a major part of my world. (One unusually bad-tempered reader once asked me to do fewer astronomical essays, but, of course, I paid no attention.)

On the other hand, I suspect the science that finds its way into my essays least often (considering its importance) is chemistry. Now you may think this is strange. After all, I received my Ph.D. in chemistry some centuries ago. (Well, it feels like centuries.) What's more, I still retain my academic title of Professor of Biochemistry at Boston University School of Medicine. So why don't I write about chemistry?

Two reasons. First, I know too much in that subject and have more trouble, therefore, making matters clear and lucid. There's always the tendency to throw in more than I need to. Second, having spent years studying and teaching the subject, I'm just a wee bit tired of it.

Imagine my surprise, then, at discovering when I began to put this book together that the last seventeen essays are quite unusual in subject matter. No less than eleven of the seventeen are primarily chemical. The remaining six are astronomical, but even two of these contain a good deal of chemistry.

This has never happened before and all I can say is that I hope you won't mind. In fact, I'm not too proud to ask a favor. Please don't mind.

# PART I
# PHYSICAL CHEMISTRY

# 1.

# MADE, NOT FOUND

I received an advertisement from a writers' magazine some time ago. They wanted me to subscribe.

Actually, that's a lost cause for them because I don't subscribe to writers' magazines, nor do I read books on how to write, nor do I take courses in the subject. The few times I have accidentally collided with such things, I have quickly learned that there are many things I do, and don't do, that are *wrong,* and that gets me nervous. Obviously, if I find out too much about what I do wrong, I will become unable to write and sell, and that would be a fate far worse than death.

So I glanced at the advertisement with lackluster eye—and was riveted at once by the fact that they had personalized it, typing in my name in an appropriate empty space. Here is what it said:

"Imagine how great these words would look in the pages of a national magazine or on the cover of a nationwide bestseller—BY ISAAC ASIMOV."

I was astonished. I don't have to imagine it. I've *seen* it.

The advertisement went on talking to me personally: "There's nothing like seeing your name in print, or the extra income that manuscript sales can bring . . . Today you have four good reasons to give free-lancing . . . another try."

*Another* try? I haven't quit the first try yet.

Clearly, the computer had not been programmed to omit established writers from its list. Or else I've got such a funny Russian name that the computer just couldn't believe I was really a writer.

Such a thing is not impossible. The Russian chemist Dmitri Ivanovich Mendeleev (1834–1907), who made what may well be the most important chemical advance of the nineteenth century, was turned down for a Nobel Prize in 1906 largely because he had a funny Russian name, instead of a sensible-sounding one in German, French, or English.

So let's start off with Mendeleev—

In 1869, Mendeleev worked out the periodic table of the elements (see "Bridging the Gaps," in *The Stars in Their Courses,* Doubleday, 1971). In this table, he arranged the elements in order of atomic weight, and did it in ranks and files in such a way that elements with similar chemical properties fell into the same row.

In order to make the arrangement work, Mendeleev was forced to leave gaps here and there; gaps which he daringly maintained would be filled by as-yet-undiscovered elements.

Thus, he left gaps underneath the elements aluminum, boron, and silicon, and called the elements that (he said) would eventually fill those gaps: "eka-aluminum," "eka-boron," and "eka-silicon." The word *eka* is Sanskrit for "one," so that the missing elements were "one below" aluminum, boron, and silicon respectively.

Mendeleev turned out to be completely right. In 1875, eka-aluminum was discovered and named "gallium"; in 1879, eka-boron was discovered and named "scandium"; and in 1885, eka-silicon was discovered and named "germanium." In each case, the properties of the new elements were precisely those predicted by Mendeleev from the regularities revealed by the periodic table.

A couple of the gaps pointed out by Mendeleev, however, were not filled in his lifetime. There were, for instance, two gaps below the element manganese. The one just below it was "eka-manganese," and the one below that was "dvi-manganese," the *dvi* being the Sanskrit word for "two," and these remained unfilled.

In 1914, seven years after the death of Mendeleev, the English physicist, Henry Gwyn-Jeffreys Moseley (1887–1915) rationalized the periodic table in terms of the new theories of atomic structure (see "The Nobel Prize That Wasn't," in *The Stars in Their Courses*). Moseley made it possible to give each element a unique "atomic number." It became obvious that if two elements had consecutive atomic numbers, then there could be no undiscovered element lying between them. Furthermore, if there were a gap in the list of atomic numbers, there had to be an undiscovered element to fill that gap.

The gaps represented by eka-manganese and dvi-manganese were still unfilled in Moseley's time, but now they had atomic numbers. Eka-manganese was element #43, and dvi-manganese was element #75, and from here on in they will be referred to by those numbers.

By Moseley's time, radioactivity had been discovered, and it seemed that all elements with atomic numbers of 84 or higher were radioactive. Elements with atomic numbers of 83 or less, however, seemed to be stable.

Suppose we ignore the radioactive elements, therefore, and confine ourselves to elements with atomic numbers of 83 or less. And suppose, further, that we refine what we mean when we talk of a "stable" element.

In 1913, the English chemist Frederick Soddy (1877–1956) demonstrated that elements could exist in a number of varieties, which he caled "isotopes." All the isotopes of a particular element fitted into the same place in the periodic table, and isotope is, in fact, from Greek words meaning "same place."

It was eventually shown that every element, without exception, possessed a number of isotopes, sometimes as many as two dozen. The isotopes of a given element differ among themselves in their nuclear structure. All the isotopes of a given element contain the same number of protons in the nucleus (a number equal to the atomic number), but have different numbers of neutrons.

As it happens, all known elements with atomic numbers of 84 and over have no stable isotopes. Every known isotope of all those elements are radioactive; some more intensely than others. Only three isotopes of atomic numbers of 84 or over are radioactive to so small an extent that an appreciable fraction of their atoms can remain unchanged over eons of time. These are uranium-238, uranium-235, and thorium-232.

The numbers attached to the isotope names represent the total number of protons and neutrons in the nucleus. Thus, uranium has an atomic number of 92, so that uranium-238 has 92 protons in the nucleus and 146 neutrons, for a total of 238. Uranium-235 has 92 protons in the nucleus and 143 neutrons. Thorium has an atomic number of 90, so thorium-232 has 90 protons in the nucleus and 142 neutrons.

As for the elements with atomic numbers of 83 and less, all those known in the time of Moseley and Soddy had one or more isotopes that were stable, and that could exist for indefinite periods without change. Tin has ten such stable isotopes: tin-112, tin-114, tin-115, tin-116, tin-117, tin-118, tin-119, tin-120, tin-122, and tin-124. Gold has only one: gold-197.

By and large only stable isotopes exist in nature, together with a few radioactive isotopes whose radioactivity is very feeble. Most of the radioactive isotopes exist only because small quantities are formed in the laboratories through nuclear reactions.

Well, then, when Moseley worked out the scheme of atomic numbers, there remained exactly four elements with atomic numbers of 83 or less that had not yet been discovered. They were elements #43, #61, #72, and #75. Chemists were confident that all four would be discovered in time and that all four were stable or (as would eventually be said) that all four had at least one stable isotope.

* * *

Element #72 lies directly under zirconium in the periodic table, so that it might be called "eka-zirconium" by Mendeleev's system. As a matter of fact (as is now known), element #72 very closely resembles zirconium in all its chemical properties. The two elements are more nearly twins than any other two elements in the periodic table.

This means that whenever zirconium is separated from other elements, advantage being taken of the way in which its chemical properties differ from those of other elements, element #72, with its properties matching those of zirconium, always separates out with it. Every sample of zirconium dealt with, prior to 1923, was always about 3 percent element #72, but chemists didn't realize that.

Two scientists, the Dutch physicist Dirk Coster (1899–1950) and the Hungarian chemist Gyorgy Hevesy (1885–1966), working in Copenhagen, made use of X-ray bombardment, which, as Moseley had shown, gave results that depended upon the atomic number of the element and not upon its chemical properties. If hafnium were present in zirconium ores, it ought to react to X-ray bombardment differently from zirconium no matter how twinlike the two elements were, chemically. Finally, in January, 1923, Coster and Hevesy detected element #72's presence in zirconium and eventually isolated #72 in quantities sufficient to study its properties.

Coster and Hevesy named element #72 "hafnium," from the Latinized name of Copenhagen, where the element was found. Hafnium, as was eventually discovered, possesses six stable isotopes: hafnium-174, hafnium-176, hafnium-177, hafnium-178, hafnium-179, and hafnium-180.

Meanwhile, three German chemists were working on elements #43 and #75 (eka- and dvi-manganese). The chemests were Walter Karl Friedrich Noddack (1893–1960), Ida Eva Tacke (1896–    ), who eventually married Noddack, and Otto Berg. They could estimate the chemical properties of these two undiscovered elements from their relationship to manganese, and they closely investigated those minerals that they felt might possess quantities of the two elements.

Finally, in June, 1925, they had sufficiently clear evidence of the presence of element #75 in a mineral called gadolinite. The next year they managed to isolate a gram of the newly discovered element and determined its chemical properties. They named it "rhenium" from the Latin name of the Rhine River in western Germany.

Rhenium as was eventually found out, had two stable isotopes: rhenium-185, and rhenium-187.

Whereas hafnium is not a particulary rare element, being considerably more common than tin, arsenic, or tungsten, and being difficult to isolate only because of its great similarity to zirconium; rhenium, on the other hand, is one of the rarest of the elements. It is only about a fifth as common as gold or platinum, so it is no wonder it was so hard to detect.

At the time they announced the discovery of rhenium, Noddack, Tacke, and Berg also announced the discovery of element #43, and named it "masurium," after a region in East Prussia, which was then part of Germany and is now part of Poland.

Here, however, the three chemists were misled by their own eagerness. Other chemists could not confirm their work and "masurium" faded out of the chemical consciousness. The announcement was premature and element #43 remained undiscovered.


As late as 1936, then, there still remained two gaps among the elements of atomic number 83 or less; these were elements #43 and #61. There were eighty-one elements known with one or more stable isotopes and, apparently, there were two to go.

Working on the problem, after the announcement of masurium had been mistaken, was an Italian physicist, Emilio Segre (1905–    ). All attempts to isolate element #43 from likely minerals, however, failed. Fortunately, though, Segre had the special advantage of having worked with the Italian physicist Enrico Fermi (1901–1954).

Fermi had grown interested in the neutron, which had first been discovered by the English physicist James Chadwick (1891–1974) in 1932. Till then, atoms had been frequently bombarded with alpha particles, which were positively charged, and which were repelled by the positively charged atomic nuclei. This increased the difficulty of carrying through nuclear reactions.

Neutrons, however, had no electric charge (hence their name) and would not be repelled by atomic nuclei. As a result, collisions took place more easily and frequently than would have been the case with alpha particles. Fermi discovered that neutrons were even more effective if they were made to pass through water or paraffin first. Under these conditions, neutrons (which usually travel at high speeds under the conditions of their initial liberation) collide with atomic nuclei of hydrogen, oxygen, or carbon, bouncing away without interaction. In the process, they give up some of their energy and, by the time they have passed through the water or paraffin, they have slowed down considerably. Such slow neutrons strike nuclei with less force and therefore have a smaller chance of bouncing off, and a greater chance of penetrating the nucleus.

When such a slow neutron enters an atomic nucleus, that nucleus frequently gives off a beta particle (which is actually a speeding electron). The nucleus loses the negative charge of the electron, which is the same as saying it gains a positive charge. *That* is the equivalent of saying that one of the neutrons in the nucleus is converted to a proton. Since the nucleus has now gained a proton, its atomic number becomes one higher than it was before.

Fermi carried through a number of neutron bombardments that converted an element to another that was one higher in atomic number, and, in 1934, it occurred to him to bombard uranium with neutrons. Unarium had the highest atomic number (92) of any known element, and Fermi thought that neutron bombardment of uranium might form element #93, which was one that was (as far as was then known) unknown in nature (see"Neutrality!" in *The Sun Shines Bright,* Doubleday, 1981). Fermi even thought he had succeeded, but the rules were too complex to make

23

that thought certain, and they led to something even more exciting (and ominous) than the creation of a new element would have been.

Segre, thinking of Fermi's work, realized that it was not necessary to go off the end of the periodic table to create a new element. If chemists couldn't find element #43, why not *form* it by bombarding molybdenum (atomic number 42) with neutrons? It would thus be made, not found.

Segre visited the University of California and discussed the matter with the American physicist Ernest Orlando Lawrence (1901–1958). Lawrence had invented the cyclotron and could carry through the most energetic subatomic bombardments in the world (at that time). For instance, Lawrence could use his cyclotron to set up an energetic beam of "deuterons," the nuclei of hydrogen-2.

The deuteron consists of a proton and neutron in rather loose association. When a deuteron approaches an atomic nucleus, the proton can be repelled and forced away from the neutron, and that neutron can then continue onward into the nucleus.

Lawrence bombarded a sample of molybdenum for several months with deuterons until the sample was highly radioactive. He then sent the sample to Segre, who had returned to Palermo, Italy, and was now working on the problem with Carlo Perrier.

Segre and Perrier analyzed the molybdenum sample and found they could isolate molybdenum, niobium, and zirconium from it and that none of these isolated elements were radioactive. However, if they added manganese or rhenium to the sample, and then separated those substances out of the sample again, the radioactivity came out with them. This seemed to mean that the radioactivity was associated with traces of magnanese or rhenium that were already in the molybdenum, *or* with some element that was so like manganese or rhenium in chemical properties as to come out with those elments.

If it were the latter case, then the element in question was very likely to be element #43, which lay between manganese and rhenium in the preiodic table. What's more, if it were element #43, it would be separated out more effectively with rhenium than with manganese, which would mean its properties were closer to rhenium than to manganese, which was to be expected of element #43.

Segre and Perrier tried to determine the properties of the new element, as best they could, by following the radioactivity as they treated their solutions in different ways. It was very difficult to do so, for they calculated they had only about a 10 billionth of a gram of element #43 as a result of deuteron bombardment of molybdenum.

In 1940, however, Segre discovered that among the products of the newly discovered process of uranium fission (stemming from the work of Fermi on the neutron bombardment of that element) was element #43. Much larger quantities could be obtained from the fission products than from bombarded molybdenum. The properties of element #43 could then be determined with considerable precision.

I might mention that I am pretty proud of myself in this connection. I wrote a story called "Super-Neutron" in February, 1941, and I managed to be right up to date. In the story, which appeared in the September, 1941, *Astonishing Stories*, I had a character talk about primitive methods for obtaining energy. He said, "I believe that they use the classical uranium fission method for power. They bombarded uranium with slow neutrons and split it up into masurium, barium, gamma rays, and more neutrons, thus establishing a cyclic process."

Right on! We science fiction writers knew about it even though the government tried to clamp a lid on the whole thing.

Notice that I call element #72 "masurium" in the story. It was the only name available, even though it wasn't legal, since Noddack, Tacke, and Berg had not truly isolated it. But then, in 1947, the German-British chemist Friedrich Adolf Paneth (1887–1958) maintained that an artificially produced element was identical in all ways with a naturally occurring one, so that the discovery of the first was equivalent to the discovery of the second.

Segre and Perrier accepted this and promptly made use of the discoverer's right of naming the discovery. They named element #43 "technetium" from the Greek word *technetos*, meaning "artificial."

Technetium was the first element to be artificially produced in the laboratory, but not the last. Nineteen more have been so produced, but technetium, of all of these, is the lowest in atomic number. Nor does it seem possible that any new element of lower atomic number will ever be artificially produced. It follows that technetium is the first artificial element, both in time and in position in the periodic table.

A study of the properties of technetium at once uncovered something unexpected. Although sixteen iso-

26

topes of technetium have been produced in the laboratory, not one of them—*not one*—is stable. All are radioactive. Nor is it conceivable, in view of what is now known, that any stable technetium isotope can be discovered in the future. Technetium, then, is the element of lowest atomic number to lack a stable isotope; it is the simplest radioactive element.

To be sure, some of the technetium isotopes are less intensely radioactive than others. The intensity of radioactivity is measured by the "half-life," which is the time it takes for half of any quantity of a substance to undergo radioactive breakdown. Technetium-92, for instance, has a half-life of 4.4 minutes, and technetium-102 has one of only 5 seconds. If the whole earth consisted of technetium-102, it would break down to a single surviving atom in less than fifteen minutes.

Technetium-99, however, has a half-life of 212,000 years; technetium-98, one of 4,200,000 years; and technetium-97, one of 2,600,000 years. In human terms these are long half-lives and if a sample of any of these were to be manufactured, very little of that sample would break down in the course of a human lifetime.

In geological terms, however, such half-lives are but trifles. Suppose that when the earth formed, 4.6 billion years ago, it consisted of nothing but one of these long-lived technetium isotopes. If that were so, then the planetary supply of technetium-99 would have been reduced to a single atom after 35 million years; technetium-98 would have been reduced to a single atom after 700 million years, and technetium-97 would have been reduced to a single atom after 430

27

million years. No conceivable amount of technetium could have survived three-fourths of a billion years, which is only 15 percent of the total existence of our planet at this time.

Any technetium that exists today in nature would only exist because of having been formed recently by the *natural* fission of uranium. The quantity so formed is excessively tiny, and it is no wonder that no chemist was ever able to locate it in any mineral, or that the announcement of Noddack, Tacke, and Berg that they had done so was in error.

Of course, when we speak of something as existing, or not existing, in nature, we are usually speaking of earth. Earth represents a vanishingly small fraction of all nature.

In 1952, the American astronomer Paul Willard Merrill (1887–1961) detected spectral lines in certain cool red dwarf stars that he identified with technetium. This has been confirmed many times over, and it has been discoverd that technetium is present in some cool stars to an amount of $1/17,000$th that of iron, which is a remarkably high concentration.

Clearly, technetium could not have been formed in such cool stars at their births and have persisted since. The half-lives of radioactive isotopes are, if anything, shortened at the temperature of the interiors of even cool stars. The technetium detected in stars, therefore, must be formed in processes that are continuing now. By trying to work out exactly what nuclear changes must exist in order to form technetium in the amounts detected, it may be that we will learn something useful about nuclear reactions in other stars. It may help us understand even our own sun a bit better.

That still leaves one element in the supposedly stable range of atomic numbers to be discussed. That is element #61, the last remaining gap in that stable range. It is one of the rare earth elements (see "The Multiplying Elements," in *The Stars in Their Courses*).

No one had ever detected element #61 in nature, although in 1926, two groups of chemists, one American and one Italian, claimed to have detected it. The former named it "illinium" (after the state of Illinois) and the latter "florentium" (after the city of Florence), both honoring the place of discovery. Both, however, proved to be mistaken.

In the 1930's, an American group bombarded neodymium (atomic number 60) with a cyclotron-produced beam of deuterons, hoping to form element #61. They probably did produce tiny traces of it, but not enough to supply definite evidence of its existence. Even so, the name "cyclonium" was suggested.

Finally, in 1945, three Americans, J. A. Marinsky, L. E. Glendenin, and C. D. Coryell, located sufficient quantities of element #61 in the fission products of uranium to study and elucidate its properties. They named it "promethium," after the Greek god Prometheus, who snatched fire from the sun for humanity, just as promethium had been snatched from the nuclear fire of fissioning uranium.

Fourteen isotopes of promethium are known and, as in the case of technetium, not a single one of these isotopes is stable. This means that there are only eighty-one elements altogether that are known to possess one or more stable isotopes, and that Noddack, Tacke, and Berg had the honor of being the very last to discover a stable element (rhenium).

Promethium is far more unstable than technetium. The longest-lived promethium isotope is promethium-145, the half-life of which is no more than 17.7 years.

Yet even 17.7 years is respectable. Two other gaps existed, in the radioactive range of elements above atomic number 83, that were not filled till after the discovery of technetium. These were elements #85 and #87. There were claims in the 1930s that they had been detected, and they were named "alabamine" and "virginium," respectively, but these claims were mistaken.

In 1940, element #85 was formed by the bombardment of bismuth (element #83) with alpha particles, and, in 1939, traces of element #87 were found among the breakdown products of uranium-235. Eventually, element #85 was named "astatine" (from a Greek work meaning "unstable") and element #87 was named "francium" (for France, the native land of the discoverer).

Astatine was unstable indeed, for its most long-lived isotope is astatine-210, which has a half-life of only 8.3 hours. Francium is more unstable still, for its most long-lived isotope is francium-233, with a half-life of but 22 minutes.

Even the elements beyond uranium, which have been formed in the laboratory since 1940, are, for the most part, less unstable than francium. Only the elements beyond atomic number 102, of which only a few isotopes are as yet known, have none with a longer half-life than francium-223.

# 2.
# SALT AND BATTERY

At a recent meeting of the Trap Door Spiders (the small and infinitely interesting little group on which I base my Black Widower mysteries), my good friend L. Sprague de Camp told the following historical anecdote, which must be true, for I never heard it before.

"Goethe," he said, "came to Vienna once to visit Beethoven, and they went out together for a walk. The Viennese, recognizing the two, were awestruck. Everyone the two great men encountered hastened to step aside and make room for them, the men bowing low and the women curtsying deeply.

"Finally, Goethe said, 'You know, Herr van Beethoven, I find these expressions of adulation quite wearying.'

"To which Beethoven replied, 'Please do not let it disturb you, Herr von Goethe, I am quite certain these expressions of adulation are meant for me.' "

The tale was greeted with general laughter, and no one laughed more heartily than I, since I am particularly fond of statements that represent artless self-appreciation (for what my readers may call "obvious reasons").

Once I had done laughing, however, I said, "You know, I think Beethoven was right. He was the greater man."

"Really?" said Sprague. "Why so, Isaac?"

"Well," I said, "one has to translate Goethe."

There was a short silence and then Jean Le Corbeiller (who teaches mathematics and is a prince of good fellows) said, "You know, Isaac, you probably don't realize it, but you have said something very profound."

Actually, of course, I did realize it, but one *must* be modest, so I said, "It's terrible, Jean. I say profound things all the time and I always fail to realize it."

You can't be more modest than that, I think.

In any case, it is quite possible that in these, my monthly essays, I may occasionally, quite by accident, say something profound. If you catch me at it, do let me know. I would appreciate it.


What I would like to talk about in this essay started with an Italian anatomist, Luigi Galvani (1737–1798). He was interested in muscle action and in electric experimentation as well. He kept a Leyden jar in his laboratory—a device that can store substantial quantities of electrical charge. When a charged Leyden jar is discharged into a person, it can give him a very nasty electric shock. Even a relatively mild discharge would cause his muscles to contract and cause him to jerk in a most amusing way (to other people, that is).

In 1791, Galvani observed that sparks from a discharging Leyden jar would, on contact with the thigh muscles of freshly dissected frogs, cause those muscles, though dead, to contract violently, as if they were alive.

That had been observed before, but Galvani went on to notice something entirely new. When a metal scalpel touched the dead thigh muscles at a time when a spark was drawn from a nearby Leyden jar, the muscle twitched even though the spark made no direct contact.

This was action at a distance. Galvani supposed the electric spark might have induced an electric charge in the metal scalpel, and that charge, in turn, might have affected the muscle.

If this were so, then perhaps one could get the same sort of action at a distance from lightning, which was, by then, known to be a discharge spark just like that of the Leyden jar, but on a vastly more enormous scale (see "The Fateful Lightning," in *The Stars in Their Courses*). If the Leyden jar could make itself felt across a few feet, the lightning should be able to do it from a distance of a few miles.

Galvani waited for a thunderstorm, therefore, and then took his frog's thigh muscles and suspended them by brass hooks from an iron railing outside his window. Sure enough, when the lightning flashed, the thigh muscles twitched. There was only one catch—when the lightning did *not* flash, the muscles twitched also.

The puzzled Galvani experimented some more and found that the twitching took place when the muscles, in contact with the brass, also made contact with the iron railings. Two dissimilar metals, in simultaneous

34

contact with the muscle, could not only produce muscle contractions, but they could do so a number of times. It seemed obvious that an electric charge had to be involved somehow, and that this charge was not permanently discharged by the contraction, but could be regenerated over and over again.

The question was: What was the source of the electricity?

To Galvani, the anatomist, it seemed that it had to be the muscle. Muscle is a very complicated substance, while iron and brass are just iron and brass. He therefore spoke of "animal electricity."

Galvani's experiments were widely publicized and the public found them exciting. After all, muscle twitching seemed to be characteristic of life. Dead muscle didn't twitch when left to itself. If it twitched under an electric discharge, it might be that electricity possessed a kind of life-force that made dead muscle momentarily act as though it were alive.

This was startling and it got some people to thinking that there might be ways to restore life to dead tissue by way of electricity. It was a great new "science fiction" notion and helped give rise to *Frankenstein,* which some people consider the first important piece of true science fiction.

To this day, a person reacting with muscle contractions to an electric shock (or to any unexpected sensation or emotion) is said to be "galvanized."


Not everyone accepted Galvani's notion of animal electricity. His chief opponent was another Italian scientist, Alessandro Volta (1745–1827). Volta thought it might be the metals that were the source of electricity and not the muscle. To test the matter, he tested two dissimilar metals in contact and, by 1794, found that these produced an electric charge even when no muscle was anywhere near.

(This embittered poor Galvani's last years. His beloved wife died and, in 1797, he lost his professorial position when he refused to swear allegiance to the new government set up by the invading General Napoleon Bonaparte. He died soon afterward in poverty and misery. Volta, on the other hand, was indifferent to governments and readily swore allegiance to anyone in power so that he prospered through Napoleon's rise to supreme power, and then prospered equally through Napoleon's fall and afterward.)

To Volta, the *fact* of an electrical charge at the juncture of two dissimilar metals was clear, though the *explanation* was not. (This is a common enough situation in science. Thus, today, the fact of biological evolution is not in dispute among sane scientists and even the general explanation is clear, but some of the details of the explanation still remain in dispute.)

Sometimes, it takes a long time for a satisfactory explanation to be reached. In the case of two-metal electricty, the proper explanation didn't arrive till a century had elapsed after the fact was first observed.

Nowadays, we know that all substances are composed of atoms, each of which, in turn, consists of a tiny positively charged nucleus at the center, and a number of negatively charged electrons at the outskirts. The positive charge on the nucleus just balances the total negative charge on the electrons, so the atom, left to itself, is electrically uncharged, or neutral.

In the case of each different kind of atom, electrons can be removed, but with different degrees of difficulty. Thus, electrons can be removed from zinc atoms with greater ease than they can be from copper atoms. To put it another way, copper atoms hold their electrons more firmly than zinc atoms do.

Well, then, imagine a piece of copper and a piece of zinc making firm contact with each other. The electrons in the zinc atoms at the metal boundary would have a tendency to slip across into the copper. Copper, with its stronger grip, wrests the electrons from zinc.

The copper, gaining negatively charged electrons, naturally gains a negative charge overall. The zinc, losing electrons, has some of the positive charge on its atomic nuclei unbalanced and therefore develops a positive charge. It is this difference in charge that can be detected by experimenters and that lends the metal combination its electrical behavior.

It might seem that the electric charge at the metal junction can be built up indefinitely as more and more electrons move from the zinc to the copper, but that is not so. As the copper develops a negative electric charge, it begins to repel the negatively charged electrons (like charges repel) and this makes it harder for more electrons to enter. On the other hand, as the zinc develops a positive electric charge, this attracts those electrons that remain in the zinc (unlike charges attract) and makes it harder for more electrons to leave.

The greater the charge the two metals develop, the

more difficult it is for a still greater charge to come into being. Very quickly, the process is brought to a complete halt at a time when only a tiny (but detectable) charge has developed.

Even this small effect has its uses. As the temperature is changed, the force attracting electrons to the nuclei of atoms is also changed, but generally by different amounts for different metals. This means that as temperature changes, the tendency for electrons to move from one metal to another across a junction, and therefore the size of the developing electric charge, will increase or decrease. Such "thermoelectric juntions" can therefore be used as thermometers.

What was in Volta's mind, however, was the development of a device from which an electric charge could be drawn off, and within which the charge could then regenerate. Since the dissimilar metals could induce a muscle twitch over and over, they should build up their electric charge over and over. If the charge is drawn off no more rapidly than it can be built up, one could have a steady flow of electricity.

This would be a great novelty for, until then, for over two thousand years, scientists had studied only "static electricity," an electric charge that is built up in a particular place and stays there and that flows, momentarily, through discharge. Volta was planning to produce "dynamic electricity," an electric charge that moves steadily through a conductor for an indefinite period. Such a phenomenon is usually called an "electric current," because in very many ways it is similar in its properties to a water current.

In order to make electricity flow, Volta needed to have something it would flow through. It was already known that electricity could be conducted by solutions of certain inorganic substances, and, in 1800, Volta used the most common of all substances—table salt, or sodium chloride.

It was his intention to begin with a bowl that was half full of salt water, and dip a strip of copper into one side of it and a strip of zinc into the other. Volta realized, however, that the effect would be multiplied if he made use of many such cups. For the purpose, he devised a number of metal strips, one end of each being zinc, the other end being copper.

Setting up a line of bowls of salt water, Volta bent each metal strip into a U-shape, dipping the zinc end into one bowl and the coper end into the next bowl. When he was done, each bowl had a zinc end in the salt water on one side, and a copper end on the other.

The total electric charge increased with the number of bowls. Volta could lead this charge from the zinc strip at one end of the series of bowls, to the copper strip at the other end, and then through the salt water of the bowls to the zinc strip from which it started. Volta thus had his electric current (which was, of course, essentially a stream of electrons, but Volta couldn't know that).

Volta called his group of bowls a "crown of cups" because they were arranged in the form of a crescent. We would call the individual cup a "cell" nowadays. Cell is a common term used for a single unit in any group of relatively small volumes, as in prisons, in monasteries, or, for that matter, in living tissue. Electricity-producing cells are sometimes called "voltaic cells'' or "galvanic cells" after the two great pioneers of the field, but they are far more commonly differentiated from other kinds of cells as, simply, "electric cells."

Another name arises from the fact that any device used to batter down something is a "battery." In Volta's time, what was usually used to batter down the wall of a city or fortress (or an opposing line of soldiers) was a "battery of artillery," or a row of cannon, sometimes lined up hub to hub, and all firing at once. Because of this the term battery has come to be used for any series of similar objects, working together to achieve some common end.

Volta's "crown of cups" is an example of this and he is the inventor of what, therefore, came to be spoken of as the "electric battery."

The term battery has come to be used so commonly for any source of electricity involving metals and chemicals (even when the source is a single chemical cell, and not a battery of them) that other meanings of the word have come to be quite subsidiary.

And since, in Volta's first battery, sodium chloride was an essential ingredient, I was inspired to give this essay the title it now has. (Why are you groaning?)


The usefulness of an electric battery such as Volta's is bound to be limited by the fact that some clumsy, or merely unwary, movement may easily upset one or more of the cups. This would not only stop the current, but would make a mess as well. It would pay, therefore, to think up some way of making a battery less splashy.

Volta managed this with another ingenious device. He prepared small disks of copper and of zinc and piled them up alternately into a cylindrical pile. Between each copper-zinc pair, he inserted cardboard disks that had been moistened with salt water. The salt water in the cardboard was enough to substitute for the half-full bowls. If the top and bottom of such a "voltaic pile" were touched by opposite ends of a wire, an electric current would flow.

Just as soon as the battery was invented, it opened new vistas in science. Only six weeks after Volta's initial report, two English experimenters, William Nicholson (1753–1815) and Anthony Carlisle (1768–1840), passed an electric current through water containing a bit of sulfuric acid to make it conductive.

They found that the electric current managed to do easily what could not be done in any other way at that time. It broke up the water molecule into its constituent elements: hydrogen and oxygen. Nicholson and Carlisle had discovered "electrolytic dissociation."

Chemists were eventually able to show by this technique that the volume of hydrogen that was evolved was twice that of oxygen. This, in turn, led to the realization that each molecule of water contained two atoms of hydrogen and one of oxygen so that the formula could be written as the now familiar $H_2O$.

Naturally, chemists wished to use electric currents to split other molecules that had hitherto defied all nonelectrical techniques. Just as, in the twentieth century, physicists raced to build larger and larger "atom smashers" in the form of particle accelerators, so in the early nineteenth century, chemists raced to build larger and larger "molecule smashers" in the form of batteries.

The winner was the English chemist Humphry Davy (1778–1829), who constructed a battery that included 250 metal plates. It was the largest up to that time and delivered the strongest electric current. He then tackled common substances such as potash and lime, which chemists of the time were convinced, contained metallic atoms in combination with oxygen. Nothing until that time could, however, pull the oxygen atoms away so as to isolate the other atoms as free meatal.

In 1807 and 1808, Davy used his battery to dissociate molecules, isolating potassium from potash, calcium from lime, and sodium, barium, and strontium from other compounds. These were all active metals, potassium being the most active of all. Potassium reacted with water, combining with the oxygen and liberating the hydrogen so energetically as to cause that gas to combine with the oxygen of the air forcefully enough to burst into flame. When Davy saw this and realized he was staring at a substance no one had ever seen before with properties no one had ever imagined, he burst into a wildly manic dance—and he had every right to do so.

In any battery, there is a substance that tends to lose electrons and become positively charged, and another substance that tends to gain electrons and become negatively charged. These are the two "electric poles," the "positive pole" and the "negative pole."

The American man for all seasons, Benjamin Franklin (1706–1790), was the first to insist that only one moving fluid was involved in electricity, and that some substances had it in excess while others suffered a deficit. He had, however, no way of knowing which substances possessed the fluid in excess and which in deficit, so, about 1750, he guessed. That decision has been universally accepted, as a convention, ever since. In Volta's copper/zinc battery, for instance, the copper (following Franklin's guess) is the positive pole and the zinc is the negative pole. If the current flows from excess to deficit, as it naturally should, then (again following Franklin's guess) it flows from the copper to the zinc.

Franklin had a fifty-fifty chance of guessing right, but he lost the gamble. The electron excess, we now know, is in the electric pole Franklin called negative, the electron deficit in the one he called positive, and the electrons (therefore, the current) flow from the zinc to the copper. It is because of Franklin's wrong guess that we are forced to say that the electron, which is the essence of the electric current, has a *negative* charge.

In designing electrical devices, it makes no difference in which direction you imagine the current to flow, as long as you are always consistent in your decision, but Franklin's wrong guess has resulted in one amusing incongruity.

The English scientist Michael Faraday (1791–1867) made use of terms suggested to him by the English scholar William Whewell (1794–1866). The two poles were "electrodes," from Greek words meaning "electrical route." The positive pole was the "anode"("upper route") and the negative pole, the "cathode" ("lower route"). This visualized the electric current flowing, as water would, from the higher position of the anode to the lower position of the cathode.

Actually, now that we follow the electron flow, the electric current is moving from the cathode to the anode, so that, if we go by the names, it is moving uphill. Fortunately, no one pays any attention whatever to the Greek meaning of the words, and scientists use these terms without the slightest feeling of incongruity. (Well, Greek scientists might smile.)

The electrons do not get consumed in the course of battery action. They can't. The electric current does not flow unless the circuit is "closed," that is, unless the electrons leave the battery at one place and then return to the battery at another place by way of an unbroken conducting route. Any time the conducting route is interrupted by something that is not conductive, such as an air gap, the current ceases.

In that case, you might think the electric current ought to flow forever, and that it could be made to do work forever, as the electrons move in eternal circles. One battery should be able to break up all the water molecules in the universe. That would mean, however, that you would have the equivalent of perpetual motion, and we now know very well that that is impossible.

In other words, the battery must eventually be used up, but why?

To see why, you must first understand that batteries of the type that Volta invented, yield an electric current through the agency of a chemical reaction. Indeed, we now know that every chemical reaction, without exception, involves the transfer (partial or complete) of electrons from some atoms to other atoms. It is the electrons, thus being transformed, that can sometimes be maneuvered through a wire and made into an electric current.

Imagine, for instance, a strip of zinc immersed in a solution of zinc sulfate. The zinc consists of neutral zinc atoms which may be symbolized as $Zn^0$. The zinc sulfate has a molecule that is symbolized as $ZnSO_4$. In zinc sulfate in solution, however, the zinc atom transfers its two most weakly held electrons to the sulfate group. The zinc therefore, missing two electrons, has a double positive charge and is symbolized as $Zn^{++}$. This is a zinc "ion," another term introduced by Faraday. Ion, from a Greek word meaning "wanderer," is an apt term because any atom or group of atoms carrying an electric charge (either positive or negative) is attracted by one electrode or the other and therefore tends to drift in that direction.

The sulfate groups gain the two electrons the zinc atoms have given up. Each has a double negative charge, therefore, and becomes a sulfate ion, or $SO_4^{--}$

Since zinc has a relatively weak hold on its electrons, particularly the two outermost, the neutral atoms in the zinc strip have a tendency to lose two electrons and slip into solution as zinc ions, leaving

45

their electrons behind in the zinc strip. The zinc strip has these electrons in excess and gains a small negative charge. The solution gains positively charged zinc ions with nothing to neutralize them and therefore has a slight positive charge. The development of these charges quickly stops any further movement of zinc from strip to solution.

Next imagine a copper strip immersed in a solution of copper sulfate. The situation is almost the same. The copper strip contains neutral copper atoms ($Cu^0$), while the copper sulfate is made up of copper ions ($Cu^{++}$) and the sulfate ions I described above. Here, though, the copper atoms have a strong hold on their electrons and the copper strip has no tendency to lose atoms to the solution. The reverse is true, for the copper ions tend to add themselves to the strip, carrying their positive charge with them. The copper strip gains a small positive charge, the solution a small negative charge, and that soon stops any further change of this sort.

Now suppose we close the circuit. Suppose we separate the two solutions not by a solid barrier, but by a porous one, through which the ions can drift under the attractive pull of one electrode or the other. Then suppose we connect the zinc strip to the copper strip by a wire.

The excess electrons in the zinc flow to the copper, which has a deficiency of electrons, so that both the negative charge on the zinc and the positive charge on the copper diminish. With both charges diminished, the zinc can continue to change from atoms to zinc ions and move into solution, while the copper ions can continue to attach themselves to the copper strip. The zinc ions, piling up in their half of the solution and making it positive, will drift across the porous barrier into the copper half of the solution, which is negative because of the loss of the positively charged copper ions.

Eventually, as the electrons continue to leave the battery at the zinc and return to it at the copper, the whole zinc strip will disappear and all the zinc will be present in solution as zinc ions. At the same time all the copper ions will disappear and be present only as neutral copper atoms in the strip. Instead of having a zinc strip in zinc sulfate and a copper strip in copper sulfate, there will be at the end only a copper strip in zinc sulfate. There will then be no more chemical change and no more electric current. Indeed, well before the chemical reaction is completely done, the electrical flow will have dwindled to the point where the battery is no longer useful.

But if batteries can be used only for a limited time and if they must then be discarded, their use can become a major expense. It might be all right for scientists, who want to run certain experiments that can be done in no other way—and hang the expense. What about the general public, however, who may want batteries for many purposes? (And we know very well all the purposes to which batteries can be put, and have been put, since Volta's day.) Is there any way in which the expense can be reduced to the point where batteries can become a practical part of every-day technology?

Obviously, there is, since even people of very moderate means use batteries constantly. I'll take that up in the next chapter.

# 3.
# CURRENT AFFAIRS

I was one of those on the speaker's platform on the first evening of the annual four-day seminar I conduct each summer, and an active, bright-eyed boy seated in the front row asked a penetrating question. As is my wont in such cases, I fixed him with my own glittering eye and said, "You're twelve years old, aren't you?"

And, as is also invariably the case, he answered, "Yes, how did you know?"

It was easy to know. As I explained once in an earlier essay, bright kids who are younger than twelve are inhibited by insecurity, while those older than twelve are inhibited by social responsibility. *At* twelve, their only aim in life is to make speakers miserable.

This twelve-year-old, whose name was Alex, was amused by my explanation. He was a likable youngster and over the next few days I much enjoyed his company. Naturally, I couldn't resist playing verbal games with him, and I didn't have it all my own way, either—don't think it.

At one point he casually mentioned his upcoming

bar mitzvah in October, so I said, "I guess you'll be turning thirteen then."

"Yes, I will be," said Alex.

"You won't be twelve anymore."

"No, I won't."

"You'll just be an ordinary dumb thirteen-year-old, eh, Alex?" I said, and smiled affectionately at him with a fatuous unawareness of the trap I had laid for myself.

Alex was aware, though. He looked up at me seriously and said, "Was that what happened to you when you turned thirteen?"

The smile was wiped from my face at once, for it was clear checkmate. All I could possibly say was a hollow, "I was an exception," to which the kid at once replied, "And so will I be."

Well, it does me good to be wiped out once in a while, and it did make a funny story, even though it was at my own expense. But it does make me just a bit less self-assured about my ability to go on with my account of the production of electricity.

Still, what choice do I have?


I ended the previous chapter by discussing a possible electric cell involving a zinc electrode in a zinc sulfate solution and a copper electrode in a copper sulfate solution—just to show the principles involved in chemical cells that produce electricity. In this particular example, however, the chemical reactions would take place so slowly that only a tiny electric current woud be produced, one that would be far too small to be of any practical use.

The easiest way of correcting this is to acidify the solution in which the electrodes rest. In effect, then, you have the zinc and copper immersed in dilute sulfuric acid. The zinc (which is much more chemically active than copper is) tends, if anything, to react with the acid too rapidly, so it is protected by a coating of inactive mercury to slow the reaction down a bit.

In the reaction, the zinc gives off zinc ions, while the copper absorbs copper ions. The essential chemical reaction is this: zinc plus copper sulfate yields zinc sulfate plus copper. In this reaction, electrons are transferred from the copper to the zinc; and from the zinc, through the circuit of wire and instruments, back to the copper.

Under these conditions, the current is strong enough to be useful and it should continue till the chemical reaction is complete and all the zinc is dissolved. Unfortunately, it doesn't. The current fades off and stops in a surprisingly short time.

The matter was taken up by an English scientist, John Frederic Daniell (1790–1845). He located the trouble. In the course of the reaction, hydrogen gas is liberated from the sulfuric acid. This hydrogen tends to accumulate at the copper electrode and insulate it so that it becomes less and less capable of taking part in the chemical reaction. As a result, the current dwindles and dies.

Daniell, therefore, attempted to make it less easy for the hydrogen to get to the copper. In 1836, he designed an electric cell in which the zinc and the sulfuric acid were inside an ox's gullet. The ox's gullet, with its contents, was then placed inside a copper container holding a solution of copper sulfate.

As a result of this, the hydrogen formed remains in the vicinity of the zinc and only slowly filters through the porus barrier of the gullet. Once outside the gullet, the hydrogen reacts with the copper sulfate, forming sulfuric acid and copper—the copper plating outside the walls of the container. The hydrogen comes through at so slow a rate, however, that appreciable amounts do not have a chance to avoid reaction with the copper sulfate and to accumulate on the copper.

Such a "Daniell cell" continues to produce electricity in sizable amounts for a prolonged period, and it was the first practical battery. (The ox gullet was quickly replaced by unglazed porcelain, which was easier to deal with and through which hydrogen could pass with equal facility.)

One disadvantage of the Daniell cell is that it has to be freshly constructed just before use. If one constructs such a cell and then allows it to stand about for a period of time before use, the materials inside and outside the unglazed porcelain gradually leak across and much or all of the chemical reaction takes place before you have a chance to make use of it.

A second disadvantage is, of course, that copper is a rather expensive material.

In 1867, a French engineer, Georges Leclanché (1839–1882), devised another kind of chemical cell, one which did not use copper. Inside the pot of unglazed porcelain, he placed a carbon rod (carbon is very cheap) and packed it around with powdered carbon and manganese dioxide. He then placed the pot in a larger container filled with a solution of ammonium chloride. He also placed a zinc rod in the container. In this 'Lechanché cell,'' electrons flowed from the zinc to the carbon.

In the course of the next twenty years, the Leclanché cell was modified by adding flour and plaster of paris to the ammonium chloride solution in order to make it a stiff paste. The unglazed porcelain was replaced by a fabric sack. The zinc rod became a zinc container in which the paste was placed, with the carbon rod and its surroundings, including the fabric, thrust into the paste. The top was sealed off with pitch and the whole enclosed by cardboard.

The result is what is usually referred to when we speak, nowadays, simply of a "battery." It is also called a "dry cell." It is not really dry, since if we were to cut it open we would find it to be moist. (It couldn't work if it were truly dry.) However, it is dry on the outside, and as long as it is intact, it can't be spilled. It can be carried in the pocket, it can be used upside down, and to the average citizen it does indeed seem to be thoroughly dry.

Sometimes it is called a "flashlight battery" because its use in flashlights was the way in which people first came in contact with it. Nowadays, of course, it comes in all sizes and shapes and is used in all the electrified games which are sold with "batteries not included" and to run all portable electronic devices from radios to computers.

In the last hundred years, a variety of different cells have been developed, each with its advantages and disadvantages, each with some uses to which it is particularly adapted. Yet even at the present time, some 90 percent of all the batteries used are Leclanché cells. It is still the workhorse.

And yet the Leclanché cell, whatever its advantages, produces electricity by oxidizing zinc or, to put it more graphically, burning zinc. Zinc is not a terribly expensive substance, but it's not terribly cheap, either. If you had to burn zinc in your furnace or in your automobile engine, you'd quickly discover you couldn't afford to stay warm in the winter or to drive your car at any time.

The only reason batteries can be used at reasonable cost is that they do jobs in which the energy requirements are low. It doesn't take much energy to run a radio, or a clock, or any of the other battery-operated gadgets.

For really high-energy requirements one must burn various "fuels," which are readily available substances that burn in air, and give off heat in the process. Fuels are usually carbon-containing substances—for example, wood, coal, and various petroleum fractions such as natural gas, gasoline, kerosene, and fuel oil.

Can one burn a fuel in a chemical cell (a "fuel cell") and get electricity out of it, instead of heat? It is possible, of course, to burn fuel in the ordinary fashion and use the heat energy to form electricity in various ways. The use of heat, however, limits the efficiency. Try as you will, if one goes from fuel to heat to electricity, one ends with, at best, only 40 to 50 percent of the available energy converted into electricity. In an electric cell, nearly 100 percent of the energy would be converted into electricity.

The first person to develop a fuel cell was an English lawyer, William Robert Grove (1811–1896), who found that he was more interested in electrical experimentation than in his legal practice.

In 1839, he devised a chemical cell that consisted of two platinum electrodes placed in dilute sulfuric acid. If that were all there was to it, of course, then there would be no chance of getting electricity out of it. With two electrodes of identical character there would be no reason for electrons to go from one to the other. Even if, for some reason, there were, platinum is a very inert metal that undergoes no chemical reactions in dilute sulfuric acid, and without chemical ractions, a chemical cell won't work.

However, while platinum is inert itself, its surface, when clean, offers a good site for chemical reactions involving other substances. Platinum is, in other words, a "catalyst," which hastens chemical reactions without itself taking any apparent part in them. This was first discovered in 1816 by Humphry Davy.

In the 1820s a German chemist, Johann Wolfgang Döbereiner (1780–1849), put this catalytic property of platinum to use. He found that when he played a jet of hydrogen on a quantity of powdered platinum, the hydrogen combined with the oxygen in the air so vigorously that it burst into flame. (Without the catalytic effect of the platinum, the hydrogen would not combine with the oxygen unless it were strongly heated.)

This was actually the first lighter of the modern type for tobacco users, and, for a time, it was popular. By 1828, some twenty thousand of these lighters were in use in Germany and in Great Britain, but since Döbereiner hadn't patented the device he never earned a penny. Besides, it proved no more than a passing fad for reasons I will explain in a moment.

Grove knew of Döbereiner's work of course, and it occurred to him that platinum might exert its catalytic effect in an electric cell as well as outside it. He therefore upended a test tube of hydrogen over one platinum electrode, and a test tube of oxygen over the other. Essentially, what he then had was a hydrogen electrode and an oxygen electrode.

Grove did obtain an electric current from this cell. He constructed fifty of them and wired them together and, in this way, obtained quite a powerful current.

This might seem a great achievement. The platinum was not used up no matter how long the cell operated. Neither was the sulfuric acid. The only change that took place within the cell was that electrons passed from the hydrogen to the oxygen, which was the equivalent, chemically, of the combination of hydrogen and oxygen to form water. This meant, of course, that the water content of the cell increased and that the sulfuric acid was constantly getting more dilute, but if water were somehow removed from the cell periodically, that would take care of that.

As a means of showing that fuel cells were possible, Grove's cell was a complete success. As a means of showing they were *practical,* however, it was a failure.

While hydrogen can be classified as a fuel, it is as inconvenient and expensive as fuel can be. It does not occur on earth as such, but must be formed by methods that consume energy.

Then, too, platinum is an exceedingly expensive substance. To be sure, the platinum is not used up in the process and is always there, but if we imagine more and more Grove cells being manufactured for a variety of uses, the capital investment in immobilized platinum grows rapidly.

Besides, although the platinum is not consumed, it is rendered useless very easily. Platinum's catalytic properties exist only if the surface is uncontaminated. Hydrogen or oxygen molecules can attach themselves temporarily to that surface and then be released after giving off or taking up electrons. There are many substances, however, that will attach themselves to the platinum surface and will then have little tendency to be released. They remain as a monomolecular film over the platinum, invisible to the eye, but preventing molecules such as those of hydrogen and oxygen from getting to the surface.

The platinum is, in that case, "poisoned" and no longer exerts its catalytic power of bringing about a combination of hydrogen and oxygen. Until such time as the platinum is removed and cleaned, the fuel cell won't work. (It is for these same reasons that Dobereiner's lighter proved impractical and was soon given up.)

It turned out to be difficult to construct a fuel cell that was practical in addition to being feasible. About 1900, another try was made by an American, W. W. Jacques. He took a number of steps in the right direction in his version.

To begin with, he did away with platinum and didn't make use of relatively expensive hydrogen. Instead he made use of a carbon rod, which could be easily formed out of coal, than which hardly anything is cheaper.

The carbon rod was placed into molten sodium hydroxide, which was, in turn, contained in an iron pot. The iron (cheapest of metals) was the other electrode. Air (not oxygen) was bubbled up past the carbon rod and, ideally, the carbon should have combined with the oxygen in the air to form carbon dioxide, thus producing an electric current. And so it did.

One might imagine that the Jacques cell would represent an irreducible minimum in expense since it is difficult to imagine coal, iron, and air being replaced by anything cheaper still. There were, however, two catches. First, the cell had to be continuously heated to keep the sodium hydroxide molten, and that meant an expenditure of energy. Second, the carbon dioxide that was formed didn't just bubble off; it combined with the relatively expensive sodium hydroxide to form the dirt-cheap sodium carbonate.

So the Jacques cell, too, was a theoretical success and a practical failure. All attempts at further modification in the direction of practicality have failed. Fuel cells do exist and can be used for highly specialized work but, to this day, none are cheap enough and practical enough to be used widely by the general public. The Leclanché dry cell remains the workhorse.

All the electric cells I've mentioned so far are used till they stop working and then they must be thrown away—unless you want to keep one for a curio or a good luck piece.

That seems sad. After all, if a chemical reaction takes place releasing an electric current in a particular direction, could one not reverse things? Could one not force an electric current through a cell in the opposite direction and in that way reverse the chemical reaction? Then, when the chemical reaction is reversed to the extent that the cell is in its original state, we could make use of it a second time, then reverse it again, and so on indefinitely.

In theory, that sounds plausible. Chemical reactions *can* be reversed if all the products of the reaction are retained and if there has not been a serious change in the state of order (that is, a too-large "entropy increase").

For instance, zinc reacts with sulfuric acid to form hydrogen and zinc sulfate. If the hydrogen is allowed to escape, a simple reversal of conditions is not going to force the zinc sulfate to change back into zinc and sulfuric acid. It needs the vanished hydrogen, too, and supplying it can be expensive.

Again, if you heat sugar and break it down into carbon and vapors, then even if you save the vapors and shake them up with the carbon you are not going to reverse the situation and obtain sugar again. The sugar breakdown represents a high degree of entropy increase and that won't lend itself to simple reversal.

We are all perfectly well aware of this. Even children, without any knowledge of entropy, rapidly gain the experience to know that some things are irreversible. Witness the following rather gruesome Mother Goose rhyme, which I think children find funny because they recognize its grotesque impossibility

*There was a man in our town, and he was wondrous wise.*
*He jumped into a bramble bush, and scratched out both his eyes.*

*And when he found that they were gone, he quick with
    might and main*
*Jumped back into the bramble bush, and scratched them
    in again.*

Yet some chemical reactions that yield an electric current *can* be reversed by an opposing electric current. In one direction of the chemical reaction to an electric current appears, as chemical energy is converted to electrical energy. If a current is forced through in the opposite direction, the original state of the cell is restored and the electricity disappears, as electrical energy is converted to chemical energy. The cell seems to accumulate and immobilize electrical energy, storing it for future use. Such a cell can be called an "accumulator," or a "storage battery."

A storage battery can be run back and forth indefinitely. It can be "discharge," converting chemical energy into electrical energy, and it can be "recharged," converting electrical energy into chemical energy, and this can be done over and over again.

Storage batteries are also called "secondary batteries" to distinguish them from objects like the ordinary dry cells and similar devices, which cannot be recharged and which are called "primary batteries." (In all honesty, I don't see why the one-use batteries are primary and the reusable ones secondary. Is it just that the former came into use first, or is there a more sensible rationale?)

In 1859, a French physicist, Gaston Planté (1834–1889), constructed the first accumulator. What he did was to take two sheets of lead with an insulating sheet of rubber between them. He then rolled the lead sheets into a spiral (lead is a soft metal) and upended the resulting spiral into dilute sulfuric acid. Since lead reacts with sulfuric acid, the acid soon came to contain lead sulfate.

Planté found that when he ran an electric current into one of the lead sheets and out the other, it produced a chemical change, and stored electrical energy in the process. From the changed sheets, he could draw an electric current until the cell was discharged, and then he could recharge it again.

Eventually, he took nine such lead double-spirals, hooked them together, enclosed the whole thing in a box, and demonstrateed that he could produce suprising quantities of electricity.

When Planté's storage battery was studied, it was found that, after being charged, one lead plate was covered with lead dioxide, the other with a spongy coating of finely divided lead.

The thing to do, then, was to start with that situation. Nowadays, the "lead-acid storage battery" consists of a number of flat gridworks of lead separated by insulators. Every other gridwork is plastered with lead dioxide, while the ones between are plastered with spongy lead. As the electric current is drawn off, both the lead dioxide and the spongy lead react with sulfuric acid to form lead sulfate and water.

However, when an electric current is forced through the battery in the opposite direction, lead and lead dioxide are formed again as the lead sulfate disappears and sulfuric acid reappears.

Lead-acid storage batteries are the familiar batteries used in automobiles and other vehicles. They supply the heavy discharge of electricity needed to start the car in the first place (after which gasoline explosions in the cylinder keep it going) and in addition, of course, a steady flow of electricity for the headlights, the radio, the power windows, the cigarette lighter, and all the other electrical equipment.

Nor does all this necessarily discharge the battery, since while the car runs, some of the energy of the burning gasoline is used to create an electric current that will serve to recharge the battery. The storage battery works for years, usually, without running down, unless you put an unusual demand upon it as, for instance, in trying over and over to start a balky car until the battery, exhausted, dies on you. Or else you might inadvertently park the car with the headlights on, and then go off and leave it in that condition for an extended period of time.

Of course, as discharge and recharge go on for month after month, gathering flaws accumulate in the plates (nothing is perfect) and eventually, the battery's ability to store electricity dwindles and it cannot be effectively recharged to more than partial capacity. You will then have trouble starting your car under even slightly difficult conditions and the battery is more likely to end up dead at inconvenient moments. The only way out, then, is to buy a new battery.

If battery charging goes on past the full capacity of the battery to store energy, the water in the sulfuric acid solution breaks up into hydrogen and oxygen, which bubble off. Little by little, the water level sinks, until the upper end of the plates is exposed. Water must therefore be added now and then to avoid such an eventuality.

There are other types of storage batteries than the

lead-acid variety. Thomas Alva Edison (1847–1931) devised a "nickel-iron battery," for instance, shortly after 1900. Still other types are "nickel-cadmium" and "silver-zinc."

The chief drawback of the lead-acid storage battery is that it is heavy. The others are all lighter, but are also more expensive and don't yield as large a slug of electricity when called upon. For that reason, the lead-acid storage battery, which was the first to be devised, is still by far the most used. There is constant talk of replacement, and some day something better will undoubtedly be found—but not quite yet, apparently.

A question arises in connection with the storage battery. Where does the electricity come from that recharges it?

The sad thing is that, in accordance with the second law of thermodynamics (otherwise known as "the general nastiness of the universe"), it always takes more electrical energy to recharge the battery than the amount of energy it will deliver on discharge.

If, then, we had to use battery electricity to recharge a storage battery, we'd be faced with a losing proposition. If, for instance, a storage battery produced as much electrical energy as five ordinary electric cells, but it took six ordinary electric cells to recharge it, then it would be better to use the five ordinary electric cells to do the work of the storage battery in each discharge cycle.

If batteries were the only source of energy, then storage batteries, in other words, would simply be a way of using up chemical cells more quickly than would otherwise be true.

Storage batteries can be of no use whatever, therefore, unless they can be charged by electricity that is produced by some means *other* and *cheaper* than chemical cells.

Fortunately, there is such a means of production of electricity, and we''ll get into that subject in the next chapter.

# 4.

# FORCING THE LINES

A few months ago, I attended a lecture on impressionistic music, which I enjoyed, because I know nothing about music, especially impressionistic music, and I find it pleasant to be educated. So I listened carefully and was particularly interested when the speaker explained that Maurice Ravel was one of the most important impressionists in music.

"Anyone who says that he left the auditorium after hearing a piece of Ravel and that he hummed the tune as he did so is kidding himself," he said, forcefully. There is no tune in the ordinary sense in Ravel's music."

I didn't say anything, of course, but I was sitting in the front row and I felt like humming at this point. And since I am utterly unselfconscious, I hummed. I didn't hum loudly, you understand, just loudly enough for the speaker to hear me.

"MMM," I hummed, "muh-muh-muh-muh-muh-muh-MMM-muh-muh-MMM-muh-muh-MMM-muh-muh-muh-MMM—" and so on.

And the speaker smiled and said, "Except in the case of the *Bolero,* of course," and everyone laughed.

For a second or so, I felt just like the nasty twelve-year-old I used to be when I was twelve years old. I loved it.

But that does show you how dangerous it is to make generalizations. That's one of the many things I try to remember in the course of writing these essays, and one of the many things I am always forgetting. So you're always welcome to hum the *Bolero* at me, figuratively speaking.

In the previous two chapters I discussed the production of electrical current by batteries; that is, by devices that convert chemical energy into electrical energy.

Well, then, could one obtain an electric current from any other kind of energy?

At the time the first batteries were constructed, there was a group of scientists or quasi scientists who called themselves "nature philosophers" and whose views ranged from honest misguidedness in many cases to down right charlatanry in some. A Danish physicist, Hans Christian Oersted (1777–1851), fell under the spell of these nature philosophers, and delivered himself of a deal of nonsense before he learned to observe more and mysticize less.

Nevertheless, it is possible to come to some useful conclusions, even if more or less by accident, from ridiculous premises, and it seemed to Oersted that there ought to be some way of interchanging elctricity and magnetism. After all, there were similarities between the two forces. Both involved attraction and repulsion, like charges (or poles) repelling each other, and unlike ones attracting each other. The force fell off with distance similarly in both, and so on.

Oersted was a scientist enough to want to demonstrate the interchangeability and not merely talk about it. He wasn't quite sure how to go about it, but one thing he thought of doing was to place a compass near a wire carrying a current to see if that current would affect the compass needle.

Toward the end of 1819, he meant to try such an experiment and, if it produced interesting results, to go on to demonstrate it in the course of a pulic lecture. He never got around to trying it out but in the course of the lecture he seemed to be carried away by his own statements and, since he had the materials on hand, he tried the experiment on impulse.

Afterward he explained why he had done what he had done, but I'm not sure I understand the explanation. My own impression is that he was caught completely by surprise by the results of the experiment and that he tried to obscure that fact.

Here's what he did. He had a strong battery by means of which he could send a current through a wire. He placed the wire over the glass of a compass, adjusting the wire so that the current would run along the north-south line of the compass needle.

When he then started the current, the compass needle suddenly jerked through an angle of 90 degrees as though, thanks to the presence of the electric current, it wanted to align itself east-west. Oersted, astonished, unhooked the wire and connected it to the battery in the opposite way, so as to reverse the direction of the current. He placed it over the needle, which had returned to north-south, and twisted again, but in the opposite direction.

The best evidence that Oersted was caught by surprise and was confused by what happened rests in the fact that he didn't follow up his experiment. He left that to others.

He did, in later life, do some reputable work in chemistry, but it was this one experiment, which he carried through without much understanding, that made him immortal. Thus, the unit of magnetic field strength was officially named the "oersted" in 1934.


Oersted announced his discovery (in Latin) in early 1820, and the physicists of Europe broke into an instant uproar of a kind that wasn't to be seen again till the discovery of uranium fission a century later.

Almost at once, a French physicist, Dominique F. J. Arago (1786–1853), showed that a current-carrying wire acted like a magnet in other ways than that of affecting a compass needle. He found that such a wire would attract nonmagnetized iron filings, just as an ordinary magnet would.

Another French physicist, André Marie Ampére (1775–1836), showed that two wires, held parallel to each other, and each carrying a current, would attract each other, magnet fashion, if the currents were flowing in the same direction in both, but repel each other if the currents were flowing in opposite directions.

Ampére arranged matters so that one wire could flip around, and then had the two carry currents in oposite directions. The one that could flip at once did

67

flip, so that the two carried currents flowing in the same direction. This is exactly analogous to the way in which the north pole of one magnet brought near the north pole of another that is free to move will cause the second to turn and present its south pole instead.

This "electromagnetism" acted very much like ordinary magnetism.

It had been known for a long time that if iron filings were sprinkled on a piece of cardboard placed over a magnet, and if the cardboard were tapped, the iron filings would fall into a pattern that made them appear to be following lines that curved from one pole of the magnet to the other pole. The English scientist Michael Faraday called them "magnetic lines of force."

Each line of force represents a curve along which the magnetic intensity has a constant value. Thus, an iron filing can slide along the curve of such a line with minimal effort. To move from one line to another requires a greater effort. (This is analogous to the way in which we can walk about on a flat flor with little effort, staying on the same "gravitational line of force," but must use more effort to move across those lines by going up or down a ramp.)

A wire with an electric current passing through it also exhibits the existence of magnetic lines of force. If the wire passes through a hole in a piece of cardboard, and iron filings are sprinkled on the cardboard, which is then tapped, the filings will align themselves in a series of closely spaced concentric circles marking out the shape of the lines of force.

Suppose, then, that a wire is twisted into a cylindrical bedspring, so to speak. Such a twisted wire is called a "solenoid," from a Greek word for "pipe," since the twists of wire seem to mark out the walls of a pipe.

If a current is passed through such a solenoid, the individual curves of wire have currents flowing in the same direction. The magnetic field of each curve reinforces those of the others, so that the solenoid is a stronger magnet than the same wire would be if it were straight, with the same current flowing through it. In fact, the solenoid resembles a magnet very much indeed, for there is a north pole at one end and a south pole at the other.

The circular lines of force, curving around the wire, combine into families of concentric ovals going up the outside of the solenoid and down the inside. Outside the solenoid these ovals, as they get larger and larger, move farther and farther away from each other, as spokes do when radiating outward from the hub of a wheel. Inside the solenoid, the ovals are forced closer and closer together as they radiate inward. The magnetic intensity rises as the lines of force are pushed closer together, so that the inside of the solenoid shows stronger magnetic properties than the outside does.

Some solid materials have the property of being able to accept an unusual number of magnetic lines of force, and of these, the most remarkable is iron, which can concentrate the lines of force enormously. (That is why iron is particularly susceptible to magnetic attraction.)

If the wires of a solenoid surround a bar of iron, the magnetic properties of the solenoid intensify further. In 1823, the English physicist William Sturgeon

(1783–1850) shellacked wire (to insulate it) and wrapped eighteen turns of wire about a bar of iron and demonstrated this.

He then used a horseshoe-shaped bar of iron weighing 7 ounces and wrapped turns of wire about it. He ran a current through the wire, and the horseshoe became a magnet that could lift 9 pounds of iron—twenty times its own weight. When Sturgeon broke the circuit and put an end to the current, the horseshoe lost its magnetic property at once and dropped the iron it held. Sturgeon had invented the "electromagnet."

In 1829, the American physicist Joseph Henry (1797–1878) heard of Sturgeon's electromagnet and thought he could do better. Clearly, the more turns of wire one made about the bar of iron, the stronger the magnet. However, when one really made many turns the wire would come into contact with itself over and over. The wire must therefore be well insulated with something more than shellac, so that the electric current would not flow through the entire mass but would patiently follow the long path of the wire round and round.

Henry decided to insulate the wire with silk, and, for the purpose, made use of his wife's silk petticoat. (I have not been able to find out what remarks his wife made when he broke the glad tidings to her.) Once he had his insulated wire, he wrapped thousands of turns about his iron bar and by 1831 had an electromagnet of no great size that could lift more than a ton of iron when the current was running—and drop it with a great clang when the current was turned off.

Not only could one convert electricity into magnetism, but one could, in this way, make magnets that were far stronger than the ordinary kind.

But could one reverse matters and make electricity out of magnetism?

One person particularly interested in this was Michael Faraday. He tried to make electricity out of magnetism four times and failed each time. In 1831 (the year Henry made his great electromagnet), Faraday set up a fifth experiment as follows:

He took an iron ring and, on one side of it, wrapped coils of wire. This wire he attached to the poles of a battery and interrupted the circuit with a key that would break the circuit if it were left open, but complete the circuit if it were pressed down and closed. By closing or opening the key, Faraday could start or stop the current passing through the coil of wires, and could in this way magnetize or demagnetize the iron ring the coil of wires enclosed.

On the other side of the ring, Faraday wrapped coils of another wire that was *not* attached to a battery. He hoped to be able to start a current flowing through it even though no battery was attached.

But how would he be able to tell whether or not this second coil of wire would gain an electric current? We cannot sense an electric current directly, and a wire with a weak electric current coursing through it looks exactly the same as a wire without one.

Here Faraday made use of an application of Oersted's original experiment. In 1829, almost immediately after Oersted's announcement, the German physicist Johann S. C. Schweigger (1779–1857) placed a magnetized needle behind glass and in front of a semicircular scale. If this device is incorporated into an electric circuit in the proper way, then, when the current flows, the needle is deflected to one side or the other (as Oersted's needle was). This device is called a "galvanometer," from Galvani, whom I mentioned in the previous chapter.

Faraday, therefore, attached a galvanometer to the second coil, and was now ready.

If he depressed the key in the first coil and started current going through it, the iron ring would become a magnet. This now magnetized iron ring passed through the second coil (the one without a battery) and, it seemed to Faraday, the magnetic ring would then start a current flowing through the second coil and that current would be registered by the galvanometer. Faraday would, in other words, have turned electricity to magnetism on one side of the ring, and magnetism back to electricity on the other side.

Faraday then closed the key, started the current, and what happened proved to be unexpected. As the current started, the galvanometer needle jerked, so that current was flowing through the second coil as Faraday had expected—but only for a moment. Though Faraday kept the key closed, the current did *not* continue. The galvanometer returned to zero and remained there. However, when the key was opened and the current in the first coil ceased, the galvanometer needle jerked briefly in the opposite direction.

In other words, current was induced in the second coil at the moment the current in the first coil was initiated and at the moment it was stopped. If conditions remained stady, with the electric current either continually present or continually absent, nothing happened.

Faraday thought of an explanation of this. When the electric current was initiated in the first coil and the iron ring became a magnet, magnetic lines of force came into being and spread outward to full expansion. In doing so, the lines of force moved across the second coil and initiated a current there. Once the lines of force reached their full expansion, they moved no more. They no longer cut across the second coil, and there was no further current. When the current stopped, however, and the iron ring ceased being a magnet, the magnetic lines of force collapsed, crossed the second coil in doing so, and again initiated a current—in the opposite direction.

Faraday concluded that to convert magnetism into electricity, one must arrange to have magnetic lines of force sweep across the wire (or across any material that can conduct electricity)—or, in reverse, have a wire (or other conductor) move across magnetic lines of force.

To demonstrate this, he set up a solenoid attached to a galvanometer, then thrust a bar magnet into its interior. When he thrust the magnet into the interior, its lines of force were cutting across the wires, and the galvanometer needle moved in one direction. When he pulled the magnet out of the interior, again its lines of force cut across the wire, and the galvanometer needle moved in the other direction. Whenever he held the magnet still at any point within the solenoid, there was no current.

There is a story that Faraday carried through this demonstration at one of his public lectures, and a woman afterward asked, "But, Mr. Faraday, of what use is this?" Faraday answered, "Madam, of what use is a newborn baby?" Another version has William E. Gladstone, then a freshman member of Parliament but eventually to be four times Prime Minister, asking the question. Faraday is supposed to have answered, "Sir, in twenty years, you will be taxing it."

I don't quite believe this story because the comparison with a newborn baby is also told of Benjamin Franklin at the rising of the first balloon, but even if it were true, I am impatient with such answers. Why must an interesting scientific demonstration be "of use"? It is sufficient that it increases our understanding of the universe, whether it is "of use" or not.

At the time Faraday worked all this out, the law of conservation of energy had still not been established as the unbreakable fundamental rule it is now considered to be. In hindsight, though, with this law in mind, we might ask where the electrical current comes from when a magnet is pushed into a solenoid. Is the magnetic energy slowly being converted into electrical energy? With every surge of electric current, does the magnet weaken slightly until finally it is a totally unmagnetic piece of iron with all its magnetic energy bled off into electricity?

The answer to that is: No!

The magnet retains its full strength. No matter how often and how continuously the magnet is pushed into the solenoid and taken out, it doesn't weaken a bit.

It can produce, in theory, an infinite number of surges of current without any loss to itself.

But surely it is impossible that we are getting something for nothing, isn't it? Absolutely! And we *aren't* getting something for nothing.

Magnetic lines of force resist being pushed across electrical conductors, and electrical conductors resist being pushed across lines of force. If we were pushing an ordinary bar of iron into a solenoid, and then pulling it out again, we would be expending some energy in order to overcome the bar's inertia. If, however, we pushed a magnetized bar of iron into a solenoid and then pulled it out again, we would have to expend an additional amount of energy because we would be forcing the lines across the wires. The same thing is true if we moved the solenoid over a magnetized bit of iron, and then lifted it off again. Once more, there would be an additional amount of energy expended, as compared with pushing it over a nonmagnet and lifting it off again, for additional energy would be required to force the solenoid across the lines.

And it is this additional energy that is converted into electrical energy.

Faraday next labored to devise some way of having a conductor cut across magnetic lines of force continuously, so that an electric current would be initiated that would flow steadily, instead of in momentary surges.

Two months after he had run the experiments that showed that an electric current could have magnetism as its source, Faraday set up a thin copper disk that could be turned on a shaft. Its outer rim passed between the poles of a strong magnet as the disk turned. As it passed between those poles, it continuously cut through magnetic lines of force so that an electric current ran continuously in the turning copper disk.

The current ran from the rim of the copper disk, where the turning motion and, therefore, the electrical pressure was highest, to the shaft, where the motion was essentially zero. If one hooked up a circuit, making sliding contact with the rim of the turning disk at one end, and with the shaft at the other, an electric current would pass through the circuit for as long as the copper disk turned.

It was still 1831, and Faraday had invented the electrical generator, or "dynamo" (from a Greek word for "power"). Naturally, this first dynamo was not very practical, but it was improved by leaps and bounds as the decades passed until continuous flows of electricity could be fed into cables, carried cross-country, and routinely brought, in any reasonable quantity, into every factory, office, and home. The little electrical outlets in the walls became a ubiquitous feature of life in the United States and other industrialized countries, and all of us, when we want an electrical apparatus to work, simply plug it into the right spot in the wall and forget it.*

The trick is to keep the copper disk (or all the later equivalents—now called "armatures") turning, for it

---

*Faraday's generator produced "direct current," flowing in one direction continually. Modern generators usually produce "alternating current," flowing in surges one way, then the opposite, changing direction sixty times a second or so—but that's a topic for another essay some day.

takes considerable energy to force it across the magnetic lines.

We can imagine such disks with cranks attached and gangs of slaves, in relays, sweating at those cranks under the encouraging strokes of long whips, but— no, thanks. Fortunately, by the time electric generators came into being, steam engines could do the turning. In that way, the energy of burning fuel could run generators and produce electricity.

It is much cheaper to burn fuel than to burn zinc or other metals, so generator electricity can be turned out in quantities that far exceed anything that can be produced by batteries. That is why when storage batteries run down, they can be profitably recharged— not by other batteries, which would have the effect of trying to lift yourself by placing your arms under your own armpits—but by generator electricity. It is also why automobile storage batteries can be recharged as you drive by the energy of burning gasoline running a small generator.

To be sure, you can turn, at best, only some 40 percent of the burning fuel into electricity, the rest being lost as heat (thanks to the good old irritating second law of thermodynamics). If you could set up an electric cell in which fuel could be made to react with oxygen, little by little, nearly 100 percent of the energy of the oxidation could be turned into electricity—but no one has yet worked out a practical "fuel cell" of this kind. And if one did, it is very unlikely that it could be built in such sizes and quantities as to compete with generator electricity.

Besides, the armature doesn't have to be turned by the action of a steam engine that burns fuel for energy. It can be turned by falling water or by wind (the same principle as the waterwheels and windmills of the pre-industrial world). Niagara Falls, for instance, is the source of a great deal of electricity that involves no burning fuel, no great heat loss, and no pollution.

In fact almost any source of energy—tides, waves, hot springs, temperature differences, nuclear power, etc.—can, in principle, be used to run a generator and produce electricity. The trick is to find practical ways of doing so on a large scale.

Considering the cheapness and huge quantity of generator electricity available, one would think that batteries would disappear altogether. Who needs the trifling bit of expensive electricity they give rise to, when you can get all you want for much less per watt by plugging into the wall.

The answer to that lies in the very phrase "plugging into the wall." You don't always want to be tied to the wall by a length of wire. You may want something transportable, a radio, wristwatch, movie camera, flashlight, or toy, and for that you need batteries. If all you want is just a small, weak current for limited tasks, for an object that you wish to be self-contained and unplugged, then a battery is what you should use.

Electricity can do some of its tasks with nonmoving parts. It is the heat generated by a current of electricity through a resistance that does all you need for lights, toasters, ovens, and so on.

For the most part, though, you want electricity to make motion possible. If you can find a way to make an electric current cause a wheel to turn, the turning of that wheel can be made to produce other types of motion.

This should be possible. In this universe, things can often be reversed. If a turning object, such as an armature, can generate an electric current, then an electric current ought to, in reverse, make an object turn.

Almost as soon as Faraday invented the electric generator, Joseph Henry reversed the process and invented the electric motor. Between the two of them, they initiated the age of electricity.

Batteries and generator electricity will both continue to be useful, and even indispensable, throughout the foreseeable future, and yet the energy source in decades to come will probably increasingly involve a totally different way of forming electricity, one that doesn't use either chemical reactions or magnetic lines of force.

I'll take that up in the next chapter.

# 5.

# ARISE, FAIR SUN!

In recent years, there have been numerous books of lists of one kind or another. And if enough people make enough lists in enough categories, it becomes inevitable that any given object should eventually be on one list or another. Even me!

Naturally, I would not be surprised to be on the list of someone's ten favorite science fiction writers. I was not prepared, however, to be found on someone's list of the ten sexiest men in America. Naturally, I know that I'm one of those ten, but I didn't really think anyone other than myself had discovered that fact.

It was not an undiluted triumph for me, though. I was placed on the list on condition that I get rid of my "silly sideburns."

Fat chance!

In the first place, I like them, and, in the second, they have unparalleled importance as a recognition device, and that is important to anyone in the public eye. I was again made aware of this, a few days ago, while having lunch in one of New York's more high-brow eating establishments.

During the lunch, a very attractive young lady approached me diffidently and asked for my autograph. I obliged in my usual suave manner and said, as I signed, "How did you know I was me?"

And she replied, "Because you *look* like you."

She meant my sideburns, of course, which are distinctive, since few people other than myself have the cast-iron self-assurance to be seen in the public eye with quite so luxuriant a set.

Yet identifying someone or something by his or her or its looks can lead to mistakes, as many have found out. After three successive chapters on different ways of producing electricity, I begin a fourth with two such misidentifications through appearance.

In the 1740s, gold mines were discovered in what was then eastern Hungary, and is now northwestern Rumania. The usual avid search uncovered more veins of gold elsewhere in Hungary, but sometimes the quantity of gold obtained from such veins was disappointingly small. Hungarian mineralogists naturally got to work in order to find out what was wrong.

One of them, Anton von Rupprecht, analyzed ore from a gold mine in 1782, and found that a nongold impurity accounted for the gold that was not obtained. Studying this impurity, Rupprecht found it had some properties that resembled those of antimony, an element well known to the chemists of the day. Judging from its appearance, therefore, he concluded that antimony was what he had.

In 1784, another Hungarian mineralogist, Franz

Joseph Müller (1740–1825), studied Rupprecht's ore and decided that the metal impurity was not antimony because it did not have some of that metal's properties. He began to wonder if he had a completely new element, but didn't dare commit himself to that. In 1796, he sent samples to the German chemist Martin Heinrich Klaproth (1743–1817), a leading authority, telling him of his suspicions that he had a new element and asking him to check the matter.

Klaproth gave it all the necessary tests and, by 1798, was able to report it as a new element. He carefully, as was proper, gave Müller (not himself or Rupprecht) credit for the discovery, and supplied it with a name. He called it "tellurium," from the Latin word for "earth" (not a very imaginative name, in my opinion).

Tellurium is a very rare element, less than half as common in the earth's crust as gold is. However, it is commonly associated with gold in ores, and since few things are as assiduously searched for as gold, tellurium is found more often than one would expect from its rareness.

Tellurium is (as was eventually understood) one of the sulfur family of elements, and the Swedish chemist Jöns Jakob Berzelius (1779–1848) was not surprised, therefore, when, in 1817, he found tellurium in the sulfuric acid being prepared in a certain factory. At least, he found an impurity that *looked* liked tellurium so that he took it for granted that that was what it was.

Berzelius was not an easy man to fool for long in this way. Working with the supposed tellurium, he found that some of its properties were *not* like those of tellurium. By February, 1818, he realized that he had still another new element on his hands, one that strongly resembled tellurium. Since tellurium was named for the earth, he named the new element for the moon, and since Selene was the Greek goddess of the moon, he called it "selenium."

In the periodic table, selenium falls between sulfur and tellurium. Selenium is not exactly a common element, but it is more common than either tellurium or gold. Selenium is, in fact, nearly as common as silver.

Selenium and tellurium were not particularly important elements for nearly a century after their discovery. Then, in 1873, there came a peculiar and completely unexpected finding. Willoughby Smith (I know nothing about him otherwise) found that selenium would conduct an electric current with much greater ease when it is exposed to light than when it is in the dark. This was the first discovery ever made of something that was eventually called "the photoelectric effect"; that is, the effect of light upon electrical phenomena.

This behavior of selenium made it possible to develop the so-called electric eye. Imagine a small evacuated glass vessel containing a selenium-coated surface that is part of an electric curcuit. A beam of light shines into the vessel so that the selenium is a conductor. An electric current passes through the selenium and acts, let us say, to keep a door closed, a door that would ordinarily be pulled open by some device.

The beam of light extends across a path along which people approach. As a person passes through the beam of light, the glass vessel is momentarily in darkness. The selenium no longer conducts a current and the door swings open. It's right out of *Arabian Nights!* You don't even have to say, "Open, sesame."

But why should light affect electrical conductivity?

Well, why not? Light and electricity are both forms of energy and, in theory, any form of energy can be transformed into any other (though not necessarily completely).

Thus, electricity can produce light. The flash of lightning during a thunderstorm is the result of an electrical discharge, and when electricity is forced across an air gap, a bright spark results. In 1879, Thomas Alva Edison in the United States and Joseph Wilson Swan (1828–1914) in Great Britain invented the incandescent electric light that is still used today and that produces light from electricity in enormous quantities.

It was, however, easy, even in Willoughby Smith's day, to see how an electric current could produce light. It was not so easy to see how light could produce an electric current.

The beginnings of an answer came in 1887, when the German physicist Heinrich Rudolf Hertz (1857–1894) was experimenting with oscillating electric currents that produced sparks across an air gap (and discovered radio waves in this fashion). Hertz found that sparks were produced more easily when light fell upon the metal points from which the sparks were being emitted. As in the case of selenium, the passage of electric current was being made easier by light, but now it seemed to be a general phenomenon, and something that was not confined to a single element.

In 1888, another German physicist, Wilhelm L. F. Hallwachs (1859–1922), sharpened things a little. He showed that a metal plate that was negatively charged tended to lose that charge when it was exposed to ultraviolet light. A metal plate that was positively charged was not affected by ultraviolet light.

Why should the two types of electric charge behave differently in this respect? In 1888, physicists could not say.

At this time, however, physicists were studying the effect produced when electricity was forced not merely across an air gap, but through a vacuum. When this happened, there was accumulating evidence that something was radiated outward from the cathode (that is, from the negatively charged portion of the circuit). This was referred to as "cathode rays," and there were arguments as to whether it consisted of lightlike radiation, or of a stream of tiny particles.

The argument was not finally settled until 1897, when the English physicist Joseph John Thomson (1856–1940) produced observations that showed quite clearly that cathode rays were a stream of tiny particles, each carrying a negative electric charge. The were *really* tiny, too. Thomson showed that they were much smaller than atoms. One of these particles is only $1/1,837$ as massive as the most common form of hydrogen atom, which is the least massive atom that exists.

The cathode ray particles were named "electrons," a name suggested six years earlier by the Irish physicist George Johnstone Stoney (1826–1911) for the minimum electric charge that could exist, assuming there was such a minimum. As it turned out, the charge on an electron *is* such a minimum under ordinary laboratory conditions. (Quarks are thought to have even smaller charges, some two-thirds that of an electron and some one-third, but quarks have never yet been detected in isolation.)

As long as physicists thought of electrons only in connection with cathode rays, they seemed to be only fundamental bits of the electric current; the "atoms of electricity," so to speak. Here, however, is where the photoelectric effect began to show its importance in the development of the great revolution in physics that took place at the turn of the century.

The German physicist Philipp E. A. Lenard (1862–1947), beginning in 1902, studied the photoelectric effect intensively. He showed that ultraviolet light, falling on various metals, brought about the ejection of electrons from their surfaces. It was this loss of electrons that carried off negative charge. If the metals were uncharged to begin with, negatively charged electrons were still ejected, leaving a positive charge behind.

The fact that electrons could be ejected from uncharged metals showed that they were not merely bits of electricity, but were components of atoms. At least, that was the simplest way of accounting for Lenard's discovery, and continuing experiments in later years confirmed the notion.

Since electrons were ejected by the photoelectric effect from a wide variety of elements, and since (as nearly as anyone could tell) all the electrons shared the same properties, whatever the element of origin might be, it semed to follow that electrons were components of *all* atoms. The difference between atoms of different elements would have to depend, at least in part, upon the number of electrons each contained, or upon their arrangement, or both, but could not depend upon the nature of the electron itself.

This sort of thinking set physicists on the track of atomic structure and, by 1930, the atom assumed the familiar picture it has had ever since. It consists of a tiny central nucleus built up of two relatively massive types of particles, protons and neutrons, the former carrying a positive electric charge equal in size to the negative charge on the electrons, and the latter uncharged. Surrounding the nucleus are a number of very light electrons.

Since the negatively charged electrons are on the outskirts of the atom and are very light and therefore easy to force into motion, while the positively charged protons are in the center of the atom and are, besides, relatively massive and, therefore, comparatively immobile, it is only the movement of the negative particles that produces the electric current. There is therefore radiation from the negative electrode, or cathode, and not from the positive electrode, or anode. And that is why ultraviolet light causes the ejection only of electrons, causing a loss of negative charge and, eventually, leaving behind a positive charge.

The picture most of us have is of the neutrons, protons, and electrons as little spheres. Actually, they must all be described in terms of quantum theory, which gives a good mathematical description but does nothing for us pictorially. There are no analogies drawn from common experience that would help us understand what these subatomic particles would "look like."

The development of the quantum theory is also bound up with the photoelectric effect.

Lenard noted that if light of a particular wavelength ejected electrons, all those electrons came off at the same speed. If the light was made more intense, more electrons were ejected, but at no greater speed. If light of shorter wavelength was used, however, electrons were ejected at greater speed, and the shorter the wavelength, the greater the speed. A dim light of short wavelength would eject few electrons, but would eject those few at high speed. An intense light of longer wavelength would eject many electrons, but at lower speed.

If the light was sufficiently long in wavelength (the "threshold wavelength"), the speed of ejection would fall to zero, and there would be no electrons ejected no matter how intense the light. The value of this threshold wavelength varied from element to element.

(For his work on the photoelectric effect, Lenard received the 1905 Nobel Prize in physics. The trauma of German defeat in World War I embittered Lenard, however, and he became notorious as one of the few notable scientists to become a convinced Nazi in the early days of that movement and to remain one throughout his life. Even in this way, he may have unwittingly served humanity, for he denounced modern theoretical physics as "Jewish" and, therefore, wrong. Since he had Hitler's ear, he may have helped persuade Hitler not to lend much support to nuclear research and thus prevented Nazi Germany from getting the nuclear bomb in time to allow it to win the war.)

Classical physics could not explain the connection between wavelength and the photoelectric effect. Something else had to be sought for, and something else existed.

In 1900, the German physicist Max K. E. L. Planck (1858–1947) had worked out the quantum theory to explain the manner in which wavelengths were distributed in the radiation from a hot body. No suitable equation based on the notion of energy as a continuous substance would work, so Planck supposed that energy came in discrete bundles he called "quanta" (Latin for "how much?"). Energy could not emerge from the hot body in amounts smaller than the quanta, but the size of the quanta varied with wavelength. As wavelength grew shorter, the size of the quanta grew correspondingly larger.

The equations based on quantum theory described the wavelength distribution in the radiation of hot bodies perfectly but, for some years, physicists (including Planck himself) considered it all merely a mathematical trick designed to solve this one problem, and didn't really think that quanta actually existed.

In 1905, however, Albert Einstein (1879–1955) showed that quantum theory explained all the puzzles involved in the photoelectric effect. One quantum of energy knocked out one electron. If the wavelength of light was too long, the quantum was too small to break the atom's grip on its electrons and there was no ejection. As the wavelength grew shorter, the quantum would get larger and eventually become just large enough to force an electron away from its atom so that it could be ejected. That would be the threshold wavelength. As the wavelength continued to grow shorter, the ejection would be brought about with greater energy and the electron would move off at greater speed. Since the atoms of different elements hold their electrons with different amounts of energy, the threshold wavelength naturally varies from one element to another.

This was the first time that quantum theory fully explained a phenomenon for which it had not been designed. It lent the theory an enormous credibility, so that Einstein deserves almost equal credit with Planck for establishing it. When, in 1921, Einstein received the Nobel Prize in physics, it was for his work on the photoelectric effect and not for his theory of relativity.

Once it is understood that light can knock electrons out of atoms, the behavior of selenium loses its mystery. Once the light knocks electrons loose, these can drift about easily and that makes a larger electric current possible

In the 1940s scientists at Bell Laboratories, notably the English-American physicist William Bradford Shockley (1910–    ), were working with substances that could conduct electricity, but only with difficulty.

They did not conduct as well as the metals, but they were not as stubbornly nonconductive as, say, sulfur, rubber, or glass. They were therefore called "semiconductors."

Certain semiconductors could be made more conductive if the substance of which they were composed was treated with small quantities of elements whose atoms contained one atom too many to fit into the crystal lattice of the semiconductor; or whose atoms contained one atom too few.

When a semiconductor contains an occasional extra electron that lacks a place in the lattice, it tends to drift, increasing the ease with which a current can flow through. Since the extra electrons add a negative charge to the semiconductor, the latter is an "n-type."

When a semiconductor is an occasional electron short, there is a hole in the lattice, and the hole tends to drift in the opposite direction that an electron would. It acts like a particle with a positive charge, and the semiconducting property is again enhanced. Such a semi-conductor is a "p-type."

Shockley and the others found that by combining n-type and p-type semiconductors in various ways, devices could be built that served the function of various radio tubes. These new devices require no vacuum, as radio tubes do, so they are "solid-state devices." When vacuums exist, they must take up considerable room to work properly, but solid-state devices don't need the room and can be very small. The latter also require no glass so they are sturdy and leak-proof; and they work at low temperatures so that they require very little energy and need no warming-up period.

In 1948, the "transistor" was developed and a new era of electronic devices was initiated.

When an n-type and a p-type semiconductor are combined there is an "n-p junction" between them. There is always a small negative charge on the electron-rich n-side of the junction and a small positive charge on the electron-poor p-side of the junction. If the n-side of such a device is connected to the p-side by a conducting wire, electrons flow from the n-side through the wire to the p-side. A very tiny current flows for a while until the electrons from the n-side fill enough holes in the p-side to stop the current.

The current was too small and short-lived to be of use, but, in 1954, the scientists at Bell Telephone discovered, by accident, that a p-n silicon junction could produce a sustained current of respectable size once it was exposed to light. It was the selenium discovery of eighty years before all over again.

The reason this happens is that the light knocks an electron out of a silicon atom and leaves a hole behind. If the device is hooked up to an electric circuit, the electron moves in the direction of the drifting electrons and into the wire. Meanwhile the hole moves in the opposite direction until it meets an incoming electron and is filled.

This current never stops as long as the light shines, for countless new loose electrons and new holes are continually being formed by the light, so that there are always new electrons to leave the device at one end and always new holes to be filled at the other.

Because such a device produces electricity, it is an electric cell just as are the chemical devices I described in the previous two chapters. Because the electricity is formed by the action of light, it is sometimes called a "photoelectric cell."

The light acts to keep one side of the cell continually rich in electrons and the other side continually poor in them. This difference in electron density produces an "electromotive force" that tends to make the electrons move in such a way as to even out the disparity. Electromotive force is measured in volts, so such a cell is sometimes called a "photovoltaic cell." When sunlight does the work of knocking electrons out of atoms, the device is called a "solar cell."

Solar cells convert the energy of sunlight directly into an electric current, and such currents are the most useful and versatile form of energy in today's world. The vision immediately arises of virtually free electricity supplied by a sun that shines endlessly—or for several billion years anyway. There are, however, catches.

1. Sunlight is copious, but it is dilute. That is, the whole world gets far more energy from the sun than it can use in the form of electricity, but one square meter of the earth's surface doesn't get much. That means we would have to spread solar cells over a large area to get the sizable quantities of electricity we would need.

2. Solar cells are not very efficient. The first photoelectric cells, those involving selenium, converted less than 1 percent of the energy of light into electricity. Later solar cells, made of silicon usually, could convert about 4 percent of light into electricity, though efficiencies of up to 20 percent are now possible. Banks of cells would have to be spread out over five to twenty-five times the area they would need if they

were 100 percent efficient. That means it would take many thousands of square miles of solar cells to supply the world with the electricity it needs.

3. Although sunlight is free, solar cells are not. Silicon is a very copious elment, the second most common in the earth's crust. it is not found as an element, however, but only in combinations with other elements. To separate elemental silicon from these combinations is difficult and, therefore, expensive. It must also be made very pure and then just the proper quantities of impurities of particular types must be added. The result is that solar cells are surprisingly expensive for their size. If you imagine thousands of square miles of them in bank upon bank, and consider the maintenance costs, the location and replacement of defective ones, the damage done by wildlife, by weather, by accident, by deliberate vandalism, it would all be the most expensive "free" energy you ever heard of.

4. Although sunlight is free, it isn't always available. There are clouds and mists and dust galore. In most of the world's most thickly inhabited areas the weather is sufficiently unsettled so that you can by no means depend upon the sunlight supply, especially in the winter when you need unusual amounts of energy for lighting and heating. Even if you switch to areas where sunlight is fairly constant and other uses for the land are largely nonexistent—such as various desert areas—it is still night half the time. What's more, even the clearest desert air scatters some light and renders it ineffective for the purpose, and this effect grows greater the farther the sun is from the zenith.

Indeed, much of the sun's energy outside the visible light region is absorbed by the atmosphere altogether.

It may, in the end, prove more effective if we simply continue to make our solar cells cheaper and more efficient and then lift the whole thing into space. Solar cells in space have, in fact, already proved useful. They have been used to power a number of satellites where the quantity of energy required is low and where other sources are difficult to arrange. I am talking now, however, of very large-scale production.

We might put a solar-power generating station, with square miles of solar cell banks, into a geostationary orbit so that it would hover over a particular spot on earth's equator more or less permanently. There would be no atomosphere around the station to interfere and scatter and absorb light. so that the entire range of radiation would be available. There would be no night to speak of, since the station would enter earth's shadow for only brief periods about the time of the equinoxes. There would be no life forms to interfere and be interfered with, and casual vandalism would not be likely. (There is possible damage by meteoroids and micrometeoroids, of course.)

A bank of solar cells in space could produce up to sixty times as much electricity as the same area of solar cells could produce on earth's surface.

Of course, electricity in space would do us no good if it stayed there, but it could be turned into microwaves and beamed to earth in a form more concentrated than sunlight. It could then be collected by relatively small banks of receiving cells that could turn it again into electricity.

There is no way of being very optimistic that the

project of solar energy production in space can be easily set up. It will surely take a long time and a great deal of labor and money, to say nothing of involving enormous personal risks for those working on it.

Still, the expense would be only a small fraction of what the nations of the world now seem glad to spend on weapons of war they dare not use; and the risk to human life is an even smaller fraction of what nations of the world seem glad to risk for the sake of their hatred and suspicions.

The possible benefits are incalculable, as the clean, cheap energy of the sun replaces that derived form the slow and expensive chemical oxidation of metals, and the dirty burning of fossil fuels.

Arise, fair sun—

# PART II
# BIOCHEMISTRY

# 6.

# POISON IN THE NEGATIVE

Yesterday I sat down to write my 321st essay for *Fantasy and Science Fiction*. I called it "How High is the Sky?" and it went swimmingly. I was pleased at the ease with which I worked out its construction. It practically wrote itself and I scarcely had to look anything up. I whistled while I worked.

And then, when I reached the last page and launched into my climactic paragraphs, I thought to myself: Why does this suddenly sound familiar to me? Have I ever written an essay like this before?

As it happens, I am widely noted as a shy and reserved person of extraordinary modesty, but if there is one thing about myself of which I'm just a weensy bit proud, it is my phenomenal memory. So I punched my recall button and up on my internal display screen came an essay called "The Figure of the Farthest" (see *Of Matters Great and Small*, Doubleday, 1975). Hoping earnestly that my memory had misfired, I looked it up. It turned out to be essay #182, first published in the December, 1973, issue. There it was. That earlier essay was essentially what I had just written.

I promptly tore up what I had spent most of the day writing, and fell into disgruntled thought. What else should I write?

For a while, I could think of nothing but subjects I had already dealt with. In fact, I was just coming to the horrifying conclusion that I had finally written everything there was to write, when my dear wife, Janet, entered my office with a concerned look on her face.

Goodness, I thought to myself, the sweet woman is so attuned to my moods that she could feel my misery, telepathically, from the other end of the apartment.

"What do you want?" I growled, lovingly.

She held out her hand. "You forgot to take your vitamins today," she said.

Ordinarily I greet a sentiment like that with an amiable snarl and a few affectionate cursory remarks. This time, however, I beamed and said, "Thank you so much, darling," and swallowed the stupid pills with a big grin.

You see, it occurred to me that I had never written an essay on vitamins.


I presume that human beings have always suffered from vitamin deficiencies, but this usually happened when they were undernourished or confined to a monotonous diet (or both)—as, for instance, if they were in prison, or in besieged cities, or were totally impoverished.

In general, they were then considered to have died of hunger or of one of the many diseases with which human beings were afflicted. Such deaths were endured stoically in the good old days, especially if the dead and dying were varlets, knaves, churls, and other members of the lower classes.

But then a brand-new peril began to strike sea voyagers—

The diet on shipboard was generally monotonous and bad. There was no refrigeration in the good old days, so there was no use storing anything on shipboard that spoiled or went moldy too easily. Consequently, the standard foods for sailors at sea were items such as hardtack and salt pork, which lasted practically forever, even at room temperature, for the good and sufficient reason that no self-respecting bacterium would touch the stuff.

Such items supplied the sailors with calories, and very little else, but sea voyaging in ancient and medieval times consisted largely of hugging the coast and making frequent stops during which sailors could get real food, so there was no problem.

Then, in the fifteenth century, came the Age of Exploration, and ships began making longer voyages during which they remained at sea for longer intervals. In 1497, the Portuguese explorer Vasco da Gama (1460–1524) circled Africa and completed the first successful voyage by sea from Portugal to India. The voyage took eleven months and, by the time India was reached, many of the crew were suffering from scurvy, a disease characterized by bleeding gums, loosened teeth, aching joints, weakness, and a tendency to bruise.

It was not an unknown disease for it was also suffered by those who were under long siege during wartime, and had been specifically remarked and commented upon since the time of the Crusades at least. This was the first occasion, however, in which the disease had appeared on shipboard.

Naturally, no one knew the cause of scurvy, any more than anyone at the time knew the cause of any disease. Nor did anyone suspect that the trouble might be dietary since the natural belief was that food was food, and if it stopped the hunger pains, that was it.

Scurvy continued to plague sea voyagers for two centuries after da Gama, and it was a serious matter. Sailors who were down with scurvy could not do their work, and the ships of early modern times were all too prone to sink in a storm even when the entire crew was able-bodied and hardworking.

And yet there were hints that scurvy could be handled.

The French explorer Jacques Cartier (1491–1557) sailed three times to North America between 1531 and 1542, exploring the Gulf of St. Lawrence and the St. Lawrence River, and laying the foundations for French dominion in what is now the Province of Quebec. On his second voyage, he wintered in Canada in 1535–36. Adding to the poor food on shipboard was the continued lack of anything else during the winter, so that twenty-five of Cartier's men died of scurvy, and nearly a hundred others were disabled to one degree or another.

According to the story, the Indians had the sufferers drink water in which pine needles had been soaked, and there was a marked improvement as a result.

Then, in 1734, an Austrian botanist, J. G. H. Kramer, was with the Austrian army during the War of the Polish Succession. He noticed that when scurvy appeared it was almost always among the rank and file of the soldiers, while the officers generally seemed to be immune. He noticed that the common soldiers lived monotonously on bread and beans, while the officers frequently had green vegetables to eat. When an officer didn't eat his green vegetables, he was liable to get scurvy just as though he were a private. Kramer recommended that fruits and vegetables be included in the diet to prevent scurvy. No one paid attention. Food was food.

Scurvy was a particular problem for Great Britain, which depended on its navy to defend its shores and protect its commerce. Clearly, if its sailors tended to be disabled by scurvy, it was quite possible that the navy might, at some crucial moment, be unable to perform.

A Scottish physician, James Lind (1716–1794), had served in the British navy, first as a surgeon's mate and then as a surgeon, between 1739 and 1748. That gave him an excellent opportunity to observe the absolutely harrowing conditions on board ships.

(Samuel Johnson, in those days, said that no one would serve on board ship who had the wit to get into jail. He said that ships, as compared to jails, had less room, worse food, worse company, and offered the chance of drowning. During wartime in the eighteenth century, the British lost about eighty-eight men to disease and desertion for every one killed in action.)

In 1747, Lind chose twelve men who were disabled with scurvy (there were, of course, plenty to choose from), divided them into groups of two, and gave each pair a different dietary supplement. One pair had two oranges and a lemon each day for the six days the supplies held out and *that* pair recovered from their illness with astonishing quickness.

Next came the task of convincing the British navy to feed the sailors citrus fruits regularly. This was almost impossible to do because, as we all know, military officers have a rigid quota of one new idea per lifetime,* and the British admirals had apparently all had theirs already, at the age of five or thereabouts.

Then, too, Captain Cook (1728–1779) during his voyages of exploration had lost only one man to scurvy. He obtained fresh vegetables at every opportunity, and he also added sauerkraut and malt to the rations. Somehow it was the sauerkraut and malt that got the credit, though they were not particularly effective, and that confused the issue.

Then the American Revolution came along, followed by the French Revolution, and the sense of crisis grew. In 1780 (the year before the climactic battle of Yorktown, when France, for one crucial moment, seized control of the western Atlantic), 2,400 British sailors, one-seventh of the total, were down with scurvy.

In 1797, the British navy was put almost entirely

*This statement offended a military officer, who sent me an annoyed letter. Of course, there are exceptions—but finding them is so hard.

out of action when the sailors, driven to despair by their inhuman treatment, rose in massive mutiny. One of the demands of the mutineers was that they be given a ration of lemon juice. Apparently, the common sailors, not surprisingly, didn't really enjoy scurvy and, even less surprisingly, had more brains than the admirals did.

The mutiny was put down by a judicious mixture of barbaric punishment and reluctant giving in. Since lemons from the Mediterranean were expensive, the British Admiralty settled on limes from the West Indies, which were not quite as effective, but were cheaper. British sailors have been called "limeys" ever since.

In this way, scurvy disappeared as a major threat on British vessels, but Lind was dead by then and could not savor the victory.

And it was a purely local victory. The use of citrus fruits did not spread and all through the nineteenth century scurvy flourished on land, especially among children who were no longer breast-fed. Though enormous advances were made in medicine during that century, that actually worked *against* the proper treatment of scurvy.

As biochemical knowledge grew, for instance, it became plain that there were three chief classes of organic foodstuffs: carbohydrates, fats, and proteins. It was recognized, at last, that food was not necessarily food, but that foods differed in nutritional quality. However, the difference seemed to rest entirely in the amount and type of protein that was present and scientists tended to look no further.

In addition, the century saw the great disvovery of

the influence of microorganisms on disease. So important was this "germ theory" and so effectively did it lead to the control of various infectious diseases, that physicians began to think of *all* disease in terms of germs and the possiblity that diet had something to do with some diseases tended to be brushed aside.

Scurvy wasn't the only disease that afflicted sailors and that could be countered by diet. In the second half of the nineteenth century, Japan was westernizing itself and was rising to the status of a great power. To that end, she worked busily to build a mondern navy.

The Japanese sailors ate white rice, fish, and vegetables and were not troubled by scurvy. However, they fell prey to a disease called "beriberi." This is from a Sri Lankese word meaning "very weak." The disease produced damage to the nerves, with the result that a person with beriberi felt weakness in his limbs and great lassitude. In the extreme, the sufferer died.

The Director General of the Japanese navy was Kanehiro Takaki, and, in the 1880s, he was greatly concerned over this matter. One-third of all the Japanese sailors were down with beriberi at any one time, but Takaki noted that the officers on board ship generally did *not* get beriberi, and that they had a less monotonous diet than the ordinary sailors did. Takaki also noted that British sailors did not suffer from the disease, and there, too, the diet was different.

In 1884, Takaki decided to produce greater variety in the diet and to add some British items to it. He replaced part of the polished rice with barley, and added meat and evaporated milk to the rations. Behold, beriberi disappeared in the Japanese navy. Takaki assumed that this was so because he had added more protein to the diet.

Again, as in the case of Lind's treatment a century before, nothing further happened. Beriberi, like scurvy, was stopped on shipboard, but, again like scurvy, beriberi continued to flourish on land. To be sure, it is comparatively easy to alter the diet of a few sailors, who can be disciplined harshly for disobedience, while it is considerably more difficult to change the diet of millions of people, especially to a more expensive one, when they can barely manage to find enough of anything at all to eat. (Even today, when the cause and cure of beriberi is precisely known, it kills 100,000 each year.)

Beriberi was endemic in the Dutch East Indies (now called Indonesia) in the nineteenth century, and the Dutch were naturally concerned over the matter.

A certain Dutch physician, Christiaan Eijkman (1858–1930), had served in Indonesia and was invalided home with malaria. He finally recovered and, in 1886, he agreed to return to Indonesia at the head of a team of physicians in order to study beriberi and to determine how best to deal with it.

Eijkman was convinced that beriberi was a germ disease, and so he took with him some chickens. He hoped to breed chickens in numbers and use them as experimental animals. He would then infect them with the disease, isolate the germ, form an antitoxin perhaps, and work out an appropriate treatment to try on human patients.

It didn't work. He could not infect the chickens, and, eventually, the bulk of the medical team returned to the Netherlands. Eijkman stayed on, however, to serve as the head of a bacteriological laboratory, and continued to work on beriberi.

Then, in 1896, quite suddenly, the chickens came down with a paralytic disease. The disease clearly affected the nerves (it was called "fowl polyneuritis" for that reason) and it seemed to the suddenly excited Eijkman that it was quite analogous to the human disease of beriberi, which was also, after all, a polyneuritis.

The chickens, Eijkman felt, had finally caught the disease. Now what he had to do was to locate the polyneuritis germ in the sick chickens, and prove it was infectious by transferring it to those that were yet well, then work out an antitoxin, and so on.

Again, nothing worked. He could not locate a germ, he could not transfer the disease, and worst of all, after about four months, the disease suddenly disappeared and all the chickens got well.

The very puzzled and disappointed Eijkman set about finding what had happened and he discovered that just before the chickens had recovered, the hospital received a new cook.

The previous cook had at some point taken to feeding the chickens with leavings from the diet fed the patients at the hospital, a diet that was heavy on polished white rice—that is, rice with the outer brownish hulls scraped off. (The reason for the polishing is that the hulls contain oils that can grow rancid on standing. The polished rice, oilfree, remains edible for long periods of time.) It was while they were being fed on these scraps that the chickens grew ill.

Then the new cook arrived and was horrified at the thought of feeding food fit for people to mere chickens. He took to feeding them on unpolished rice, complete with hulls. That's when they got better.

Eijkman realized, then, that beriberi was caused and cured by diet and was *not* a germ disease. There had to be something in the rice that caused the disease, and something in the hulls that cured it. It wasn't anything that occurred in substantial quantities since the carbohydrate, fat, and protein of rice were in themselves harmless. It had to be some very minute "trace" constituent.

Trace constituents capable of sickening and even killing people were, of course, known. They were called poisons, and Eijkman decided that there was a poison of some sort in the white rice. In the rice hulls, he thought, there was something that neutralized the poison.

This was rather the reverse of the truth, but the notion of trace substances in food that produced or cured sickness proved uncommonly fruitful. Whereas Lind's and Takaki's work was important, but produced no further consequences, Eijkman's work produced a blizzard of subsequent experimentation and brought about an enormous revolution in the science of nutrition.

It was for this reason that Eijkman was awarded a share in the 1929 Nobel Prize in physiology and medicine, for by that time the seminal nature of his work was abundantly recognized. Unfortunately, he was too ill by that time to go to Stockholm to collect the award in person, and the next year he died, but, unlike Lind, he had lived long enough to witness his own victory.

Eijkman returned to the Netherlands soon after he had made his great discovery, but a co-worker, Gerrit Grijns (1865–1944), remained in Indonesia. It was he who first announced the correct interpretation. In 1901 (the first year of the twentieth century), he presented arguments for believing that something in the rice hulls did not serve to neutralize a toxin, *but was itself essential to human life.*

White rice resulted in disease, in other words, not because it possessed a small quantity of a poison, but because it *lacked* a small quantity of something vital. Beriberi was not merely a dietary disease; it was a dietary deficiency disease.

This was revolutionary thought! For thousands of years, people had been well aware that one could die through the presence of a bit of poison. Now, for the first time, they had to get used to the thought that death could result from the *absence* of a bit of something. That ''something'' was the opposite of a poison, and since its absence meant death, it was a poison in the negative, so to speak.

Once this fact was absorbed, it seemed likely that beriberi wasn't the only dietary deficiency disease. Scurvy was an obvvious example of another. In 1906, the English biochemist Frederick Gowland Hopkins (1861–1947) suggested that rickets, too, was a dietary deficiency disease. He was particularly successful in publicizing the concept and pursuading the medical profession to accept it, so he shared the 1929 Nobel Prize with Eijkman.

In 1912, the Polish biochemist Casimir Funk (1884–

1967) suggested pellagra as a fourth dietary deficiency disease.


Nutritionists naturally grew nervous over the business of some trace substances in food that represented life or death to an organism, including the human being. That was made to order for mysticism. What had to be done was to isolate the materials, determine exactly what they were, and find out how they worked. That would reduce matters to ordinary, prosaic biochemistry.

It was not enough, in other words, to work with food and to say, "Lemon juice prevents scurvy and brown rice prevents beriberi." That might be enough for people who would otherwise get those diseases, but it would not be enough for scientists.

The person who took the first step toward moving beyond the foods themselves was the American biochemist Elmer Verner McCollum (1879–1967). In 1907, he was working on the nutrition of cattle, varying the nature of the diets and analyzing the excreta. There was, however, so much food and excreta involved, and everything was so slow, that McCollum grew frustrated and weary. He decided that one had to work with smaller animals and more of them, so that studies could be made more quickly. The knowledge thus gained could be applied to larger animals— as Eijkman had done with his chickens.

McCollum moved beyond chickens. He established the first colony of white rats intended for nutritional studies, a technique the rest of the field was quick to adopt.

McCollum, furthermore, tried to break down foods into various components, sugar, starch, fat, protein, and feed these, separately and in combination, to the white rats, observing when their growth proceeded normally, and when it slowed, or when abnormal symptoms of any sort appeared.

In 1913, for instance, he showed that when he used certain purified diets, on which rats did not grow normnally, normal growth could be resumed if a little butter fat or egg-yolk fat were added. Nor was it the fat alone that did the trick, for when lard or olive oil was added to the diet, growth was *not* resumed.

It had to be some trace substance present in some fats but not in others. The next year, McCollum reported that he could extract the trace substance from butter, by using various chemical procedures, and add it to olive oil. Thereafter, that olive oil could support growth if it were added to the rats' diet.

This offered strong support to the notion of trace substances necessary to life, and deprived it of any mystical aura. Whatever the trace was, it had to be a chemical substance, and one that could be dealt with by chemical methods.

It happens that living tissue is mostly water. In this watery medium, there are solid structures made up of inorganic material (bones, for instance) or large insoluble molecules (cartilage, for instance). In addition, there are small organic molecules, many of which are soluble in water and exist in solution in consequence.

Some tissue molecules, however, are *not* soluble in water. The chief of these are the various fats and oils, which clump together separately from the water. Certain other molecules that are not soluble in water dissolve in the fat instead.

Thus, we can group the small molecules in living tissue as either "water-soluble" or "fat-soluble." Water-soluble substances in tissue can be soaked out in more water. Fat-soluble substances in tissue can be soaked out by making use of solvents such as ether or chloroform.

The trace substance essential to growth that was present in some fats and not others is clearly fat-soluble. McCollum could show, on the other hand, that whatever it was in rice hulls that prevented beriberi could be extracted with water and was therefore water-soluble. That was, in itself, conclusive proof that there was not one overall trace substance that permitted normal growth and prevented disease, but that there were at least two.

In the absence of any knowledge of the structure of these substances, McCollum had to use a simple code to distinguish them. By 1915, he was speaking of them as "fat-soluble A" and "water-soluble B" (giving his own discovery priority out of natural egocentrism).

That started the fashion of using letters of the alphabet to identify these necessary trace substances, a habit that continued for a quarter of a century until their chemical structure was known well enough for them to receive other names. Even now, however, the letter designations are frequently used, not only by the lay public, but even by biochemists and nutritionists.

Meanwhile, though, another attempt at naming had been made. Funk, whom I mentioned earlier, was working in London on these trace substances. His

113

chemical analyses had convinced him, in 1912, that whatever the trace substance was that prevented beri-beri, it contained as part of its chemical structure an atom grouping consisting of a nitrogen atom and two hydrogen atoms ($-NH_2$). This grouping is chemically related to ammonia ($NH_3$) and is therefore called an "amine" by chemists. Funk turned out to be right in this conclusion.

Funk then went on to speculate that if there were more than one of these trace substances, then all were probably one kind of amine or another. (He was wrong in this.) For that reason he called the trace substances, as a group, "vitamines"; that is (from the Latin), "life amines."

It didn't take many years for evidence to accumulate that some trace substances necessary to life did not have an amine group as part of their chemical structure and that "vitamine" was consequently a misnomer. There are many cases of this sort in science and, often, the misnomer must remain, if it has become too embedded in scientific writing and too ground into customary use to be given up. ("Oxygen" is a misnomer, for instance, and has been known to be such for nearly two centuries, but what can we do?)

In 1920, however, the English biochemist Jack Cecil Drummond (1891–1952) suggested that the final *e* of the word might at least be dropped, so that the "amine" reference would not be so overwhelmingly prominent. The suggestion was quickly adopted, and the trace substances have been known as "vitamins" ever since.

For that reason, "fat-soluble A" and "water-soluble B'' came to be known as "vitamin A" and "vitamin B''—and I will carry on the story of what we can now call vitamins in the next chapter.

# 7.
# TRACING THE TRACES

My father was a man of decided opinions. Lacking a formal education outside his vast learning in Hebrew and in biblical law and theology, he had to rely on common sense. That frequently misled him, of course, but, as I learned early in life, once he had formed his opinion, he never, by any chance, changed it—except when he happened, by some accident, to be correct in the first place.

I remember once, when my father was inveighing against the iniquities of "playing the numbers," as part of his shrewd way of seeing to it that his hopeful son would never fall prey to the wickedness and folly of gambling. (He never did.)

I listened for a while and then thought I would short-circuit it a bit and said, "I know, Poppa. You're supposed to pick a combination of three digits and there are a thousand such combinations. Therefore, your chance of picking the right combination is one in a thousand, but you are only paid six hundred to one if you win. That means that if you play a thousand numbers at a dollar a number, your chances are that you'll win once. You'll have spent a thousand dollars to get six hundred and the people who run the game keep the other four hundred."

My father said, "The chances are *less* than one in a thousand."

"No. Poppa. Suppose you take a thousand people and each picks a different combination of digits from zero zero zero to nine-nine-nine. One of them will win. so the chances are one in a thousand."

"Aha," said my father, "My smart son makes an argument! That's if every person picks a *different* combination. But who says they'll pick different combinations? They'll pick any combination they want and what if *no one* picks the right combination? That makes it less than one in a thousand."

"No, Poppa. That possibility is just balanced by the fact that in a few cases, two people will both pick the right combination."

My father stared at me with disbelief. *"Two pick the right combination? Impossible!"* And that ended the argument.

Of course, the ins and outs of probability are not always easy to follow even for trained mathematicians.

I recall another incident that came about after I began my course in quantitative analysis. I explained to my father the nature of a chemical balance and the extreme delicacy of its workings. It could weigh to a fraction of a milligram if it were properly calibrated and it its swings were properly observed—and a milligram was only about a thirty-thousandth of an ounce.

My father shook his head. "That's ridiculous," he

said. "Who could bother weighing such small amounts? They don't matter. A thirty-thousandth of an ounce of *anything* can't be important."

Nothing I said could convince him of the importance of extreme delicacy in analytical procedures.

And that brings me back to vitamins, the subject of the previous chapter.


I ended the previous chapter with the naming of two trace factors (substances necessary to life in very small amounts) as vitamin A and vitamin B; vitamin A being fat-soluble and vitamin B being water-soluble. Since every substance in the body that is soluble at all, is soluble either in fat or in water, it would be neat if there were one vitamin for each and no more. However, it's too much to hope for to have things quite that simple.

Thus, vitamin B will prevent the disease of beriberi, or cure it quickly if it already exists. It won't, however, do a thing for scurvy. There is something in orange juice that will prevent or cure scurvy, but won't do a thing for beriberi. The trace factor in orange juice was named "vitamin C" by Drummond (who had suggested the change from "vitamine" to "vitamin").

Although vitamin C, like vitamin B, was known to be water-soluble, the two had to be different somehow, for they prevented and cured two different diseases, and neither had any effect on the other's disease.

Then, in 1922, a group of nutritionists at Johns Hopkins University showed that one could prevent the bone disease of rickets, or cure it, by proper dietary methods. Certain foods must therefore contain still another trace factor, which was named "vitamin D." This, like vitamin A, was fat-soluble, but again the two had to be different in some ways, for they affected different diseases.

Vitamins were frustrating substances becasue they could be looked upon as "mysterious." If a given food that was known to contain a particular vitamin were separated into its componenets and these were chemically purified, it would be found that none of the compounds would affect the disease so that none was the vitamin, even though the compounds added up to 100 percent of the food, as nearly as could be measured. Either the vitamin was something immaterial and who knows what, or else it was an ordinary chemcial compound, but was present only in minute traces.

Naturally, if there is the slightest possiblity that something vital to health is "mysterious," we know that all sorts of nonsense will be used to victimize the general public. Since vitamins were clearly too important to be allowed to drop into mystic mumbo-jumbo, there was considerable pressure on biochemists to identify the vitamins as particular compounds, no more mysterious in nature than any other compounds—to trace the traces, in other words.

But how does one do that? Suppose one takes orange juice and adds a certain chemical that will attach itself to certain molecules in the orange juice to form an insoluble substance, while leaving other molecules untouched and still in solution. Separate the insoluble substance from the solution and then ask yourself, Is the vitamin C in the insoluble substance or in what is left of the juice?

How can you tell? The sure way is to place living things on diets known to contain no vitamin C so they will develop scurvy. Then, once scurvy appears, add to some of the diets the insoluble substance and to other diets what is left of the juice, and see which one (if either) will cure the scurvy. That one will contain vitamin C.

That's not as easy as it sounds. Scurvy can be made to appear in human beings, particularly in babies, but you can't very well experiment with babies, inducing and curing scurvy in them. You have to use some other animal and obtain the necessary information.

It turned out, unfortunately, that animals generally are far less sensitive to scurvy formation than human beings are. Diets that would give us scurvy handily, don't bother them.

By 1919, however, two types of animals were found that could be made to suffer from scurvy. One included the various types of monkeys, who are apparently close enough to us in the evolutionary tree to react as we do to the absence or presence of vitamin C. The trouble there is that monkeys are expensive animals and hard to handle.

Luckily, it turned out that the guinea pig could also be used for the purpose and would develop scurvy even more easily than the human being. What's more, guinea pigs are cheap and easy to handle.

By means of "animal assays," then, one could determine which foods had vitamin C, and which had not. One could even determine how much vitamin C a particular food had (in arbitrary units). One could also determine, in this way that vitamin C was easily destroyed by heating or by oxygen.

Most important of all, one could treat vitamin C sources chemically and follow the vitamin C content of the various fractions into which the food item was divided. Inevitably, some fractions were prepared that had vitamin C in greater concentration than any natural food had.

By 1929, the American biochemist Charles Glen King (1896–    ) and his associates had produced a solid material such that a gram of it would contain as much vitamin C as 2 liters of lemon juice, (or, to put it another way, 1 ounce of it would contain as much vitamin C as about 60 quarts of lemon juice).

In England, meanwhile, a Hungarian biochemist, Albert Szent-Györgyi (born in 1893 and still actively engaged in research today in his nineties) was investigating "oxidation-reduction reactions." In living tissue, some compounds are prone to give up a pair of hydrogen atoms (this being equivalent to "oxidation") and others are prone to accept a pair of hydrogen atoms (equivalent to "reduction").

One can picture certain compounds as capable of assisting such reactions if they themselves have a particularly easy ability to do both. Such compounds will pick up two hydrogen atoms from molecule A and give them to molecule B. They are then ready to pick up two more hydrogen atoms, pass them along, and so on. These compounds are called "hydrogen carriers."

Since oxidation-reduction reactions are vital to the

function of living organisms, it is clear that hydrogen carriers can be very important, and are worth investigating.

In 1928, Szent-Györgyi isolated a particularly active hydrogen carrier from adrenal glands. From its chemical reactions, it seemed to be related to sugars, but it had an acid group at one end of the molecule rather than an alcohol group. Such sugar-related molecules were known to biochemists and were lumped together as "uronic acids." There are a number of varieties of such uronic acids possible, however, and all that Szent-Györgyi could tell about his compound at first was that it had six carbon atoms in the molecule. He called it "hexuronic acid," therefore, *hex* being the Greek word for "six."

Meanwhile, King, working on his concentrated vitamin C material, finally managed, in 1931, to obtain a pure crystalline substance from it that showed extremely strong vitamin activity. Half a milligram ($1/57,000$ of an ounce) of those crystals added to the daily diet each day would protect a guinea pig from scurvy. There seemed no question but that the crystals were vitamin C itself. The trace had been traced and the vitamin became a definite, known material substance.

As those crystals were studied, it became clear that the were the same coumpound that Szent-Görgyi had called hexuronic acid. It seems, then, that Szent-Györgyi was the firt person to isolate vitamin C and King was the first to recognize the fact that it *was* vitamin C. As a result, the two generally share the credit of the discovery.

In 1933, Szent-Györgyi suggested that his hexuronic acid be renamed "ascorbic acid," now that its vitamin nature was understood. The new name comes from Greek words meaning "no scurvy," and that has been its name ever since, though vitamin C is still commonly used by the public as well.

Once sizable quantitities of pure ascorbic acid could be isolated (especially after Szent-Györgyi found that red peppers were particularly rich in it, and used that as a source) chemists quickly worked out its exact chemical structure, accurately placing each of the twenty atoms (six carbon atoms, eight hydrogen atoms, and six oxygen atoms) in its molecule.

Even before the exact structure had finally been worked out, methods were discovered for synthesizing ascorbic acid. The synthetic ascorbic acid is just as effective a vitamin as the natural material is. The two molecules are identical. Whether a chemist or a plant makes it, all the atoms are in the right place, and there is no way of distinguishing between them. After that, ascorbic acid could be made by the ton if that were necessary.

The isolation, structural determination, and synthesis of ascorbic acid were enough to remove any "mystery" from the vitamins. Ascorbic acid is a molecule like other molecules, made up of atoms like other atoms, and amenable to study and manipulation by the ordinary rules of chemistry. The mere fact that one vitamin could be reduced to prosaic chemistry made it reasonably certain that all could.

And all have. Every known vitamin has by now had its molecular structure worked out.

* * *

Naturally, biochemists were working on vitamin B as well as on vitamin C, and, in some ways, vitamin B was the easier task. For one thing, vitamin B, whatever it was, turned out to have a molecule that was tougher than that of vitamin C. Vitamin B was less likely to be degraded by heat or oxygen than vitamin C was, so that the former could be knocked about by the various chemical procedures used to isolate it without its suffering too much damage.

What's more, most animals are quite sensitive to the lack of vitamin B, as compared with the relatively few that are sensitive to the lack of vitamin C. It was illness in chickens, as I mentioned in the previous chapter, that gave the crucial clue to the nature of the prevention and cure of human beriberi. Consequently, animal assays of vitamin B were easier to handle than those of vitamin C, with the even more convenient white rat used in place of the guinea pig.

As early as 1912, Funk managed to obtain, from yeast, a crude mixture of crystals that tested out by animal assay to be quite concentrated in vitamin B activity.

It was because Funk detected the presence of an amine group in the vitamin B concentate, and because of his guess that all vitamins might contain them, that he invented the name "vitamine," as described in the previous chapter. And it was because, for one thing, no amine group could be detected in concentrates of vitamin C, that the final *e* was knocked off the name.

By 1926, concentrates of vitamin B were prepared that seemed to be fairly pure. Attempts at analyzing the small quantities of such concentrates (by methods that required the weighing of very tiny quantities, despite my father's skepticism as to the worth of such things) produced preliminary judgments that the molecule of vitamin B contained carbon, hydrogen, oxygen (almost all organic molecules do), and nitrogen (which a sizable quantity of them do). So far, so good, but biochemists kept plugging away, trying to get the vitamin B concentates purer, and to isolate them in greater quantities.

In 1932, a Japanese biochemist, S. Ohdake, working with tiny quantities of vitamin B material, reported the detection of sulfur atoms in the molecule. This wasn't exactly unprecedented, for sulfur atoms are to be found, for instance, in almost all protein molecules. However, of the five types of atoms most likely to be present in the molecules of living tissue— carbon, hydrogen, oxygen, nitrogen, and sulfur— sulfur is the least common. So startling was this discovery (which was soon confirmed) that vitamin B was given the name "thiamine," or, even more commonly nowadays, "thiamin." The *thi-* prefix comes from the Greek *theion,* meaning "sulfur."

Finally, in 1934, the American chemist Robert Runnels Williams (1886-1965) and his co-workers managed to refine the purification method to the point of getting completely pure thiamin. By his methods, it took a ton of rice polishings—the outer husks of the unpolished grains—to yield 5 grams ($^3/_{17}$ of an ounce) of thiamin.

The structure of vitamin B was then worked out in detail to the exact position of every atom in its molecule. In order to check whether the determination was really correct, Williams began with simpler compounds of known composition and then put them together step by step by means of chemical reactions that produced known changes. Eventually, he formed a compound that should have been the thiamin molecule, if the analysis had been correct. And, indeed, the synthetic compound proved to be the thiamin molecule, for it had the same chemical properties as the natural substance had, and it also had the same preventive and curative effect on beriberi.

The thiamin molecule contains two rings of atoms, connected by a one-atom bridge. There are also two small side-chains of atoms attached to each ring. It is the rings themselves, however, to which I want to call attention.

Rings of atoms are very common in organic compounds, and are most often made up of five or of six atoms. Very often, all five or all six atoms of the ring are carbon atoms, but sometimes one or two of the atoms in the ring may be nitrogen or oxygen or sulfur. Any ring containing atoms other than carbon is said to be "heterocyclic," *hetero-* being from a Greek word meaning "other" or "different."

Both rings in the thiamin molecule are heterocyclic. One of the rings consists of six atoms, two of which are nitrogen. The other ring consists of five atoms, of which one is nitrogen and one sulfur.

In the process of attempting to concentrate vitamin B, biochemists discoverd that they sometimes got fractions that seemed to be important to nutrition and yet had no effect as far as beriberi was concerned.

There is, for instance, a disease called "pellagra," characterized most visibly by a dry, cracked skin, that

126

occurs under conditions of limited and monotonous diets and that can be cured by broadening the diet. The dietary connection was definitely domonstrated in 1915, by an Austrian-American physician, Joseph Goldberger (1874–1929).

By that time, enough was known about vitamins for a search to begin at once for purified fractions with anti-pellagra action. At first, it seemed that substances that would cure beriberi might also cure pellagra, but the fractions being tested were sufficiently impure for it to be possible that more than one vitamin might be present.

Then, in 1926, it was found that it was possible to heat concentrates strongly enough to destroy the anti-beriberi acion, but to leave the anti-pellagra effect untouched. This made it look as though there were two vitamins, one of which had a molecule more heat-resistant (and, therefore, probably simpler) than the other.

In 1937, an American biochemist, Conrad Arnold Elvehjem (1901–1962), followed a trail of investigation that led him to try a rather simple substance on dogs suffering from "blacktongue," a disease very similar to human pellagra. A single, tiny dose sufficed for rapid and marked improvement. It was the vitamin.

Its molecules consisted of a single ring of six atoms (five carbon and one nitrogen) with hydrogen atoms and a small one-carbon acid group attached. It had first been isolated from living tissue in 1912, without any suspicion of its vitamin nature, of course. It had been formed in the laboratory, however, as long before as 1867 by a chemist named C. Huber.

Huber began with nicotine, the well-known alkaloid found in tobacco. The nicotine molecule consists of two heterocyclic rings, one of five atoms and one of six atoms. One atom of one ring was bound to an atom of the other. Huber treated the nicotine in such a way as to break up the five-atom ring, leaving only the carbon atom that was attached to the six-atom ring, and converting that atom into an acid group. He therefore named the six-atom ring, with its acid group side-chain, "nicotinic acid," as an indication of the more complicated compound from which he had obtained it.

When an organic compound is substantially changed, there is no necessary connection between the properties of the original and the product; none at all. Nicotine is a highly toxic substance. Nicotinic acid is relatively harmless. In fact, in tiny quantities, it is essential to life. It was nicotinic acid that Elvehjem had dmeonstrated to be the anti-pellagra vitamin.

That set up a problem for the medical profession. The general public was not expected to understand the finer points of organic chemistry. If nicotinic acid were hailed as a vitamin, then some people were going to think that there must be something healthy in nicotine and they were going to start smoking, or increase the dose if they were already smoking, on the assumption that this would keep them from developing pellagra.

Physicians, therefore, insisted on making use of a shortened form of "nicotinic acid vitamin." They used the first two letters of the first word, the first two of the second word, and the last two of the third, and behold, they had "niacin," which is now the most common name of the vitamin.

The same procedures that isolated concentrates containing thiamin and niacin also produced small quantities of other substances vital to life. In some cases, nutritionists knew of no diseases corresponding to deficiencies of those factors, because the substances were so widely spread in organisms and were required in such small quantitites that almost any human diet was bound to have enough of such materials for a person to get along on.

Nutritrionists and biochemists had to feed rats, or other experimental animals, on special purified diets containing only known vitamins and minerals and no other trace substances, and then, when some abnormaility about the animal developed, they had to find some food that would correct the abnormality, and search for a compound within the food that would be the vitamin.

Eventually, it became quite certain that in extracting vitamin B from food, one pulled out a whole family of somewhat related compounds—all water-soluble, all containing heterocyclic rings of one kind or another, all vital to life in tiny quantities.

The whole can be referred to as the "B-vitamin complex." Before the nature of these molecules was determined, they were named vitamin $B_1$, vitamin $B_2$, and so on up to vitamin $B_{14}$.

Most of them turned out to be false alarms, but vitamin $B_1$ is thiamin, of course. Vitamin $B_2$ is now better known as "riboflavin," vitamin $B_6$ is "pyrodoxin," and vitamin $B_{12}$ is "cyanocobalamin." Niacin doesn't have a vitamin B name, and neither do such other members of the complex as biotin, folic acid, and pantothenic acid. In fact, the only member of the B-vitamin complex in which the vitamin B name is more common than the chemical name is vitamin $B_{12}$, perhaps because the chemical name is so complicated and because it was the last to be given.

Not all vitamins belong to the B complex, of course. Vitamin C doesn't, even though it is water-soluble, because it is so different in structure from members of the complex. Vitamin C lacks nitrogen atoms in its molecule, while all memembers of the complex possess them.

Then, too, any vitamin that is fat-soluble is, by that fact alone, not a member of the B complex. Besides, fat-soluble vitamins also lack nitrogen atoms. In addition to vitamin A and vitamin D, the fat-soluble vitamins include vitamin E and vitamin K.

(What happened to the letters between E and K? Well, vitamin F was a false alarm, vitamin G was eventually identified with riboflavin and vitamin H with biotin, so those both turned out to be members of the B complex. As for vitamin K, that was named out of alphabetical order because it was involved with the mechanism of blood coagulation and in German coagulation is spelled *Koagulation*. Since the discoverers of the vitamin were German, vitamin K seemed a natural.)

Now that all the vitamins are structurally well known and can be systhesized in one way or another, there is a much diminished danger of vitamin shortage in any society that has the synthetics available. You can eat whatever pleases you and add to it a judicious selection of vitamin pills, and you will be safe from scurvy, beriberi, pellagra, and the rest.

To be sure, there are people who believe in "megavitamin therapy," *mega* being a Greek word meaning "very large." The feeling among such people is that, although tiny doses of this vitamin or that are sufficient to keep off visible disease, those diseases represent major breakdowns. Larger quantities of the vitamin might be needed to keep things moving completely smoothly, thus preventing minor disorders that, although not visible to the unaided eye, take their toll as the years pass. There is also the feeling that, though a reasonably healthy person need only take small doses of vitamins, there are disorders that are not generally recognized as vitamin-related that will benefit from large doses of some vitamin

The best-known aspect of megavitamin therapy is the use of large doses of vitamin C. This is backed by the famous American chemist Linus Pauling (1901– ) and vitamin C has been supposed to be useful in preventing colds, and even in ameliorating cancer.

I tend to be dubious about such claims. The body doesn't seem to store water-soluble vitamins, so that any supply over and above immediate needs is excreted through the kidneys.* I see no great need to consume pills in large numbers merely in order to enrich the urine.

---

*Dr. Pauling has written to say that this is not true in the case of vitamin C, and , of course, he may be right.

131

With fat-soluble vitamins, the case is different. Reactions in fat aren't as rapid as those in water, so that fat-soluble vitamins are not so easily and quickly disposed of. They are stored and tend to accumulate.

If the accumulation passes beyond a certain point, the results can be harmful. Vitamin A and vitamin D will, in large doses, produce toxic effects.

Fish and fish-eating animals may accumulate vitamins A and D far beyond the safe capacity of other animals. That's the reason that before vitamin pills became available, some youngsters had their lives made into living hells by being regularly dosed with cod-liver oil.

There are also horror tales, concerning the truth of which I am not very certain, that polar bears have livers that accumulate vitamin A to such an extent that Arctic explorers who had, on occasion, eaten polar bear livers, went on to die of vitamin A toxicity.

In the next chapter, I will take up what I consider the most unusual vitamin of them all.

# 8.
# THE GOBLIN ELEMENT

I have my faults. (Yes, I do. I insist!)

For instance, I am incredibly provinicial in some ways. Although I am a pronounced Anglophile, I simply cannot get used to British spelling and pronunciation. I have heard them say "eevolution" and "deefecate," with long *e*'s, on radio and television, and I invariably shout "evolution" and "defecate" at them, with short *e*'s, and they never listen. I've heard them pronounce schedule "shedule" and glacier "glassier" and my turning purple doesn't help.

I brood about it. To me, "colour" and "honour" and "labour" are ridiculous. All those words clearly rhyme with "flour" and that's not the way you pronounce them. And I won't even mention "gaol," which has a hard *g* if ever a word had it.

Sometimes I get so uncharitable about it that I feel myself on the point of announcing publicly that if the British can't spell and pronounce the American language properly, they ought to make up a language of their own.

Just now, in fact, I finally said so—out loud—to myself, because there was no one else around.

I wanted to find out when the word "anemia" first came to be used in medicine and so I turned to a book I have in my library, and looked up anemia—a-n-e-m-i-a. It wasn't there. There was absolutely nothing between "androsterone" and "anencephalic," and I was flabbergasted. Anemia is a very common medical term and the book purported to deal with the origin of medical terms. How could it miss?

I turned away with a muttered "dear me." I think I may even have said something stronger, like "good gravy."

And then a little light went on in my skull. I got the book again and looked at the title page. It was put out by an American publisher but the compiler of the book was a Canadian. Aha! I looked up "anaemia," and there it was.

That book will never know how close it came to being thrown out. It was only the thought that it had been invaluable to me on a number of occasions in the past that kept it on my shelves.

The word anemia, you see, comes from a Greek expression meaning "no blood." The *a-* prefix (*an-* before a vowel) as a Greek general negation meaning "no," "not," and so on. like the Anglo-Saxon *un-* and the Latin *non-*. The rest of the word comes from the Greek *haima*, meaning "blood," with the *ai* diphthong pronounced like a long *i*.

The Romans used their own diphthong *ae* (also pronounced like a long *i*) in place of the Greek *ai*. Since English uses the Roman spelling, we think of the

Greek word as *haema,* and the derived word became "anaemia" rather than "anaimia."

In English, however, the pronunciation of the *ae* became a long *e*. It seems to me, then, that the changed pronunciation makes a simple *e* sufficient for the spelling, and that's why we write "anemia." The British, however, continue to write "anaemia." For similar reasons, we write "hemoglobin," "hemorrhage," "hematology," "hemophilia," and "hemorrhoid," while they put a extra *a* into every one of those. Since heaven is just, I'm sure it sides with us in this matter.

To be sure, when this essay appears in Great Britain, the spelling will be changed to suit themselves, but I refuse to be held responsible for any consequences that may befall them as a result.

Apparently, the word anemia first came into medical usage in 1829 to describe various conditons in which there seemed to be a deficiency of blood, or, at least, of the red coloring matter of blood, so that the victim is unusually pale.

The red coloring matter of blood is "hemoglobin" and it is contained in the red corpuscles. Hemoglobin contains iron atoms, and iron atoms are not easy things to come by in food. The body conserves its iron well, however, so mostly that is not a problem. If one loses blood through accident, or through the agency of some thoughtful enemy, one naturally has trouble replacing the iron.

Young women have a particular problem because they lose blood each month when they menstruate, and it is they who most frequently suffer from "iron-deficiency anemia."

There are various other causes of anemia, however, since the manufacture of red corpuscles can go wrong in any of a number of ways, even when the iron supply is adequate. Some types of anemia are more likely to have serious consequences than others.

This brings us to a British physician named Thomas Addison (1793–1860). He is best remembered today because, in 1855, he identified a serious disease marked by atrophy of the cortex of the adrenal gland. This disease, resulting from insufficiency of adrenal cortical hormones, is still known as "Addison's disease" today.

Before that, however, in 1849, he carefully described a form of anemia that seemed particularly serious and particlularly resistant to treatment. For a while, it was known as "Addison's anemia," and as it was studied further, it began to be clear that a diagnosis of Addison's anemia was a sentence of death. All treatments failed and the victim invariably died. The disease therefore came to be called "pernicious anemia," since pernicious, from the Latin, has the meaning of "to the death."

Once the twentieth century arrived, however, physicians became aware of vitamins (as I described in the previous two chapters) and any noninfectious disease became suspect. A search began for some dietary deficiency that might account for pernicious anemia. The first hint at advance, however, came about through indirection.

An American physician, George Hoyt Whipple (1878–1976), was primarily interested in bile pigments, which are compounds that originate through the breakdown of hemoglobin.

The hemoglobin molecule contains a nonprotein portion called "heme," which consists of a large ring made up of four smaller rings, with an iron atom at the center. The body gets rid of heme, when this is desirable, by breaking the large ring and excising the iron atom for future use. The broken ring, which is the bile pigment, is then disposed of.

It occurred to Whipple that he might understand bile pigments better if he understood the details of the hemoglobin life-cycle. He therefore began, in 1917, to bleed dogs until they were decidedly anemic and then to try out various diets to see which would most rapidly lead to the rebuilding of the normal red corpuscle count.

Whipple found that a diet in which liver was prominent was more potent than any other in hastening the replacement of heme and of red blood corpuscles. In hindsight, this is not really surprising. Liver is very much the chemical factory of the body so that it is rich in vitamins and minerals (including iron). If anything is going to help in a purely nutritional way, liver is a likely choice.

Whipple was not working on pernicious anemia, but some thought his results might just possibly be useful in that direction.

Pernicious anemia had its very puzzling aspects. It might be a vitamin-deficiency disease, but if so, why do so few people get it? When someone was suffering from the disease, why was there often nothing remarkably imbalanced about his diet? And why did others with similar diets *not* necessarily get the disease?

The normal human being produces strong hydrochloric acid as part of the stomach's digestive secretions. The result is that the stomach's "gastric juice" is by far the most acid fluid in the body, and this helps **with** the process of digestion. (So acid is the gastric juice that biochemists have a difficult problem explaining how the stomach lining can endure such a constant acid bath—and sometimes it doesn't, as anyone with a stomach ulcer can testify.)

Oddly, though, the pernicious anemia victim invariably lacks hydrochloric acid, and this gives rise to the thought that there might be a disorder of digestion or absorption involved in the disease. It might be that even though the vitamin is present in the food, the victim is unable to make use of it. In that case, he might only be helped by an unusually large supply of the vitamin, so that while most would be wasted, some, by sheer force, would leak through.

So must have reasoned an American physician, George Richards Minot (1885–1950), and his coworker, William Parry Murphy (1892–    ). In 1924, Minot was so impressed by Whipple's announcement of the efficacy of liver on anemic dogs, that he decided to try a liver diet on his pernicious anemia patients. He had absolutely nothing to lose.

He started feeding them liver in large quantities, and it worked! The pernicious anemia was halted and his patients not only stopped getting worse, they started getting better.

As a result, Whipple, Minot, and Murphy all

shared the 1934 Nobel Prize for physiology and medicine. After all, pernicious anemia was no longer a sentence of death.

The suspicion that there was both an external vitamin and some internal incapacity was elevated to a strong likelihood in 1936 by the work of the American physician William Bosworth Castle (1897–     ). He showed that there had to be an "intrinsic factor," which aided the absorption of the vitamin.

We now know that the intrinsic factor is a glycoprotein (a protein molecule including a complicated sugarlike component) that must combine with the vitamin before it is absorbed. It is the lack of the intrinsic factor that is the real trouble, for the vitamin is required (as it eventually turned out) in extraordinarily small amounts. Even if that small amount were not available in the diet, which is unlikely, bacteria in the intestines could form it in ample quantities (as they can form some other vitamins, too). Indeed, the feces of untreated pernicious anemia patients are rich in the very vitamin for want of which they are dying.

There is an important catch to the liver treatment. It worked, yes, but it was a life sentence to liver eating, and in substantial quantities, too. This was better than dying, one might suppose, but as time went on, it is understandable that patients couldn't help but begin to wonder if liver wasn't a fate worse than death.

If the treatment was to be endurable, the vitamin would have to be extracted from the liver.

The American biochemist Edwin Joseph Cohn (1892–1953) tackled the problem, but he labored under a great difficulty. Whenever he divided some liver preparation into two portions through chemical treatment, the only way he could tell whether the vitamin was in one portion or in the other was to try both on pernicious anemia patients and see which helped. In every case, it took a long time to decide definitely whether a particular fraction was helping or not.

Even so, in six years of labor, from 1926 to 1932, Cohn was able to prepare a liver extract that was very efficient in alleviating pernicious anemia. Relatively small quantities of the extract would do the trick and those patients who had the extract available to them were freed from the necessity of gobbling liver day after day.

Nevertheless, Cohn did not isolate the vitamin itself. That fell to the lot of the American chemist Karl August Folkers (1906–       ).

In 1948, he and his co-workers made the key discovery that certain bacteria required the pernicious anemia vitamin for growth. A vitamin has a certain role in the chemical machinery of a cell, and its absence causes many things to go awry. Some disorders are more noticeable than others and we naturally focus on the noticeable. In the case of the human being, the most noticeable disorder following inadequate use of the pernicious anemia vitamin involves the formation of red cells. Still, just because a bacterium doesn't have red cells doesn't mean it doesn't need the pernicious anemia vitamin for other reasons. If it can make its own, fine; but if it can't, the vitamin must be provided in the nutrient mixture supporting the bacterial culture. If the vitamin is not provided, bacterial growth stops.

Folkers had found a bacterium that would grow only in the presence of the vitamin and that meant that whenever vitamin concentrates from liver (or from anywhere else) were further fractionated, the location of the vitamin could be quickly determined by bacterial assays without ever having to bother the poor pernicious anemia victims. More and more concentrated preparations were obtained and, before the year was out, red crystals were isolated that were the vitamin itself: vitamin $B_{12}$, as it was called.

There were several astonishing facts concerning $B_{12}$ that were determined once the vitamin could be dealt with directly. With respect to the daily requirement, it was the least of the B vitamins.

The daily requirement of the various B vitamins was in the milligram range. An adult male needs 20 milligrams of niacin per day, 2 milligrams of pyridoxin, 1.7 milligrams of riboflavin, 1.4 milligrams of thiamin, and so on. To put it another way, if you had an ounce of niacin and an ounce of thiamin and if you helped yourself to your daily need each day, the niacin would last for nearly four years, and the thiamin would last for fifty-five years.

The recommended daily dose of $B_{12}$, however, is about 5 micrograms for the adult male. A microgram is a thousandth of a milligram. If you had an ounce of $B_{12}$, you would have enough for yourself for 15,523 years (assuming, of course that it didn't deteriorate on standing). That would be a lifetime supply for about 220 people. Under those circumstances, it might seem astonishing that there would be a shortage.

But then, there is a second unusual thing about $B_{12}$. The molecule is surprisingly large. It is built up, if

141

my count is correct, of 181 atoms and has a molecular weight of 1,358. That makes it roughly four times as large as the other B vitamins.

In fact, it is among the largest one-piece molecules in living tissue, and here you must understand what I mean by "one-piece."

There are larger molecules by far in cells—starch, proteins, nucleic acids, rubber, and so on. What's more, chemists can form huge molecules in the laboratory—fibers, plastics, and so on. In all cases, however, these giant molecules, with molecular weights in the tens and hundreds of thousands, are made up of strings of relatively small units—the units all being similar, or even identical—and the strings are easily broken apart in to single units. Such giant molecules are "polymers."

$B_{12}$, however, is not a polymer. It can be broken up into fragments, but these fragments are unlike each other. It is therefore one piece.

Starch, protein, and nucleic acid molecules, when present in food, are too large to be absorbed as such and made use of. However, those molecules are easily split apart ("digested") into their small units. The units can then be absorbed into the body and there be put together again into giant molecules. This is not possible for $B_{12}$. It must be absorbed in one piece and its size makes that difficult. It needs an intrinsic factor, which combines with it and, so to speak, pulls it in. Without that factor—pernicious anemia.

The large size and the intricate structure of $B_{12}$ made it very difficult to work out the details. It was not till eight years after its isolation that its exact structural formula was elucidated, and that victory

142

was attained by an English biochemnist, Dorothy Crowfoot Hodgkin (1910–     ).

Her specialty was working with X-ray diffraction patterns, which are produced when X-rays bounce off atoms. If the molecules in a preparation exist in random orientation, the X-rays bounce off in random directions and if the resultant beam impinges upon a photographic film there is a central dark spot on the negative surrounded by a haze that fades off symmetrically in all directions.

If, however, a crystal is used, the molecules within it are arranged regularly, so that the constituent atoms appear in regularly repeated patterns (like those on wallpaper). The X-rays bounce off each of the repititons, in the same direction, each bounce reinforcing the next. As a result, the photographic flm will show a series of dots in particular symmetrical positions.

From the nature of the symmetry and from the separation of the dots, conclusions can be drawn concerning the position of various atoms within the molecule and, with that as a lead, the structure can be worked out. Naturally, the more complicated the structure, the more complicated the diffraction pattern, and the more difficult it is to work out the molecular structure.

Hodgkin has worked on the X-ray diffraction pattern of penicillin, for instance, and used a computer to help solve the problem. This was the first use of a comptuer in connection with biochemistry.

She then went on to $B_{12}$ and again made the use of a computer. After considerable unremitting work, she solved the problem completely and, in 1956, announced the precise structure of $B_{12}$. For this she received the Nobel Prize in chemistry in 1964.

In order to understand the structure of $B_{12}$, let's go back to heme. As I said earlier, the molecule of heme is made up of a large ring, made up of four small rings. The small rings are of five atoms each (four carbon atoms and a nitrogen atom) and these small rings are attached to each other by one-carbon bridges. The result is what is called a "porphyrin ring."

The porphyrin ring, though apparently large and unwieldy, is a very stable arrangement of atoms and occurs commonly in nature. There are many varieties of molecules containing such a ring, since small atom combinations ("side-chains") can be attached here and there to the ring. Each different side-chain, or each different arrangement of side chains, produces a new compound.

When a porphyrin with the proper side-chains in the proper arrangement contains an iron atom in the center of the ring, the result is heme, an essential component of hemoglobin. We couldn't live without it.

Many forms of life don't have hemoglobin, but they must have iron-porphyrins just the same, for these are also present in compounds called "cytochromes." Cytochromes make it possible for cells to make use of molecular oxygen in extracting usable energy from organic molecules. All cells that make use of oxygen (the vast majority of all cells that exist) must have cytochromes.

When a porphyrin ring with a somewhat different set of side-chains has a magnesium atom at the center, it is chlorophyll rather than heme. Chlorophyll is a universal component of all green plants (which are green because of the chlorophyll they contain). It is chlorophyll that makes it possible for plants to utilize the energy of sunlight in such a way as to manufacture complex organic compounds. The entire animal world (ourselves included) depends for its energy supply on the organic compounds thus built by plants.

Magnesium-porphyrin compounds are, therefore, as essential to the vast majority of all cells as iron-porphyrins are.

$B_{12}$ has a molecule that is built about a ring system that is *almost* a porphyrin. The ring system is made up of the four smaller rings of five atoms each, but there are only three one-carbon bridges connecting the small rings. The fourth bridge is missing, so that two of the rings connect with each other directly. The result is a lopsided "corrin ring."

The corrin ring has side-chains, some quite complicated, at almost every available atom. What is most surprising, however, is the central atom. It is not iron and it is not magnesium. At this point, then, let's shift to another part of the story.

Several centuries ago, copper miners in Germany were occasionally annoyed at finding a blue rock that looked as though it were malachite, an ore that yielded copper, but wasn't. This other blue rock, teated as malachite, would yield no copper and, indeed sometimes yielded vapors that made the miners sick. (The ore contained arsenic, it was eventually found out.)

The miners came to a natural conclusion. The blue rock was copper ore that had been enchanted by a spirit with a warped sense of humor. There were mischievous earth spirits in German folklore called "kobolds." (This is equivalent to the goblins of English folklore, and, indeed, both "kobold" and "goblin" may trace back to the Greek *kobalus*.) The miners, therefore, called the false ore "kobold."

This ore was investigated by the Swedish chemist Georg Brandt (1694–1768), and, in 1742, he extracted a metal from it that was not copper. On the contrary, it resembled iron a good bit, even to the point of being attracted (weakly) by a magnet. It was not iron, however, since, for one thing, it did not form a reddish brown rust.

Brandt kept the name for it that the German miners had given it, but it had gained a slightly different spelling—"cobalt." Because of this name, which it has kept ever since, cobalt can fairly be called the "goblin element" if one wants to be dramatic, and, for the titles of these chapters, I sometimes like to be dramatic.

Cobalt has come to be very useful in forming many alloys, but has it any function in living tissue?

In general, living tissue is mostly water, but if the water is removed, the dry material remaining behind can be analyzed. It turns out that carbon makes up about half the weight of the dry material.

This is as it should be. All "organic compounds" (so called because they were originally associated with living organisms) are made up of molecules contain-

146

ing carbon atoms in combination with oxygen and hydrogen and, frequently, nitrogen. These four types of atoms, taken together, make up about 88.5 percent of the dry material of mammalian tissue.

There is also a little sulfur and phosphorus in proteins, a lot of calcium and phosphorus in bones, sodium and chlorine ions dissolved in the body fluids, a bit of magnesium here and there, and, of course, iron in the red blood cells and cytochromes.

Add all these together and you end up with well over 99 percent of the weight of dry matter. It is then easy to dismiss the rest as meaningless.

However, once vitamins came into the consciousness of biochemists, they realized the importance of trace components. Might not some elements be necesary to life in trace quantities, too? If so, that less than 1 percent of the dry weight might include tiny quantities of elements that were, nevertheless, essential to life.

One way of checking for trace elements in tissue is to dry it and burn it thoroughly, leaving behind a small quantity of ash that can be analyzed. Small quantities of a variety of elements are invariably found, but that raises an important question. Are the elements there because they are part of important, even vital, molecules, or are they there just because there is always a small amount of contaminating matter in food?

When we eat, we are bound to pick up some of every element there is. Undoubtedly there are a few atoms of gold wandering about in our body, but that doesn't mean that gold is an essential component of living tissue and, as far as we know, it isn't.

The presence of an "essential trace element" becomes more likely if it is always present in all ash derived from tissue. It becomes even more likely if an animal is kept on a diet chemically free of that elment and appears to suffer as a result. The best evidence of all, however, is to find that the element in quesiton is an essential part of a molecule that is known to be necessary to life in trace amounts.

In the middle 1920s, cobalt was being found in the ash derived from living tissue, but, for ten years or so, this was dismissed as just a case of contamination.

In 1934, however, animal nutritionists were concerned with a disease that produced anemia in sheep in various regions of the world. The addition of iron compounds to the feed did not help.

But then an iron-free preparation from a mineral called limonite was found to do the tirck. The preparation was carefully analyzed and its various components, in pure form, were added to the sheep feed, one at a time. Before long, it turned out that pure cobalt chloride, added to the feed in small quantities, would cure the disease. It seemed that cobalt might be essential to life for a sheep and, it was later discovered, for cattle, too.

However, sheep and cattle are ruminants and it might be that cobalt is only useful in that special case and is not needed by nonruminant organisms such as human beings.

But then came the news, after the structure of $B_{12}$ had been worked out, that at the center of its corrin ring was a cobalt atom and that the $B_{12}$ molecule would not work without that atom. Since organisms cannot stay alive without $B_{12}$, it follows that cobalt, though present in excessively tiny traces, is essential to life.

To the cobalt, by the way, is attached a cyanide group that is, however, too tightly held to do us harm, and is present in quantities too small to do us harm even if it were not tightly held. For that reason, $B_{12}$ is now called "cyanocobalamin."

The question of how anything can be needed in such small quantities, and yet not be dispensed with altogether, will be taken up in the next chapter.

# 9.
# A LITTLE LEAVEN

My beautiful, blond-haired, blue-eyed daughter, Robyn, who is now on the job as a psychiatric social worker, got together with her lovely co-worker the other day, and decided to compose a fiery memo denouncing some practice or other they considered heinous.

They got paper and pens (the easy part) and then started brooding over the wording. Minutes passed, and nothing came to them except a dozen false starts. Finally, Robyn, throwing down her pen in exasperation, said, "Do you believe I'm my father's daughter?"

When she told me the story that night, I laughed, for when she was a little girl, there was a widespread disbelief over that very matter. Since Robyn's mother was, in the matter, completely above suspicion (either by me or by anyone else), the general theory was that Robyn had been accidentally switched in the hospital with my true offspring. (Actually, I know that is not true for Robyn has, with time, developed unmistakable Asimovian features and, if it is possible for a gorgeous woman to look like me, she does.)

Nevertheless, friends of mine staring at a little girl with blond hair who looked precisely like the John Tenniel illustrations of Alice in *Alice in Wonderland* (she was asked to play the role, at sight, in her grammar school) and then looking at me with a certain shudder of revulsion, would say, "Are you sure you weren't given the wrong child at the hospital?"

At which I would invariably put my arms around her protectively, and say, "Who cares? I'm keeping this one."

I told Robyn of this when we talked about the unwritten memo, and said that, listening to all the comments of this sort, she was in a good position to make much of the very common fantasy of children that their parents were not really their parents but that they were, instead, the kidnapped offspring of royalty.

"Never!" said Robyn, forcefully. "Never! Not for one moment at any time did I ever doubt that you and Mamma were my parents."

Which pleases me. Both Robyn and I have a strong sense of duty. I would discharge my paternal obligations punctiliously even if I didn't particularly like her, and she would be equally punctilious, I am quite certain, about being filial under such circumstances. However, there is a tight bond of affection between the two of us that makes all that duty bit an unbelievable pleasure.

And the same, I can't help thinking, goes for my science essays. Having agreed to provide the *F & SF* with one essay an issue, I would certainly perform that chore dutifully even if it proved to be a royal pain in the whatever. However, I enjoy the process so much that I keep it up month after month with a light laugh on my lips. In fact, if I have a difficulty, it lies in confining myself to doing merely twelve a year.*

I have spent the three previous chapters on vitamins, and it will seem to you probably, as you read, that I am changing the subject—but only apparently, as you will eventually see.

People discovered, in prehistoric times, that if ears of grains were heated and moistened to form a dough, and then pounded flat, a nutritious substance could be produced in quantity. It would keep so well that, if the grain were deliberately cultivated, the resulting hard bread would help support an unprecedentedly high density of population. Of course, the eating of such hardtack required good teeth, a strong digestion, and a sturdy disregard for the more effete pleasures of gourmet dining.

And then, in ancient Egypt, perhaps about 3500 B.C., it was discovered that a particular strain of wheat separated so easily from its chaff that it did not require much in the way of preliminary heating. Such unheated wheat, if ground, moistened, and beaten, would occasionally not remain flat and hard but would begin to puff up all by itself.

Undoubtedly the tendency would be to throw away

*So if you should happen to be moved to write to the Noble Editor in order to urge him to put out an issue now and then that includes *two* of my essays, go ahead. See if I care.

such spoiled material but, under the pressure of a grain shortage, such puffy material might have been baked anyway, and the result would be a soft bread, spongy and filled with tiny holes, and unparalleled in flavor and texture.

What happened (as we now know) was that yeast cells, which are always floating in air along with uncounted varieties of other spores and seeds of microorganisms, fungi, and plants, get into mashed up grain. There they live on the components of the grain, forming carbon dioxide and alcohol in the process.

If the grain has been too strongly heated, it is too hot for the yeast to live in it. If it is then moistened, pounded flat and heated to dryness it becomes too dry for the yeast to live in it. If the grain is anything other than wheat, then even if yeast lives in it, the bubbles that form escape from the grain and leave little mark, or at best, a crumbly mess. It is only in wheat, not heated strongly, and allowed to stand, that the carbon dioxide and the alcohol vapor can't escape. They are trapped in a sticky protein named "gluten," which, when baked, expands without breading, forming small gluten bubbles filled with gas. Eventually, the baking process kills the yeast and drives off the carbon dioxide and the alchohol vapor, but leaves the bubbles, now filled with air.

At first, perhaps, bread makers had to rely on each batch of dough happening to accumulate yeast. But then they stumbled upon a way of assuring success. They would just save a bit of the bubbling dough, unbaked, and bake the rest. the bit of unbaked dough they would add to a fresh batch of dough and, on standing, all the fresh dough would bubble. You could continue to pass the stuff from batch to batch, and get good raised bread all the time.

One word in English for the material that causes the dough to bubble is "leaven," from a Latin word meaning "to rise," because the dough, as it traps carbon dioxide, rises. Leaven may also be referred to as a "ferment," from a Latin word meaning "to boil," since the bubble formation is reminiscent of that which takes place when a liquid boils. The word yeast itself seems to be related to Greek and Sanskrit words meaning "to boil."

In ancient times, leaven was not thought of as being alive, since it didn't seem to have the attributes of life. It didn't move and jump around, for instance. And yet, with the wisdom of hindsight, we wonder that it didn't strike anyone as odd that a small batch of leaven could infect a whole new batch of dough over and over. Wasn't the leaven multiplying itself? And isn't that an attribute of life?

Perhaps people simply didn't wonder about such things, or if they did, in the context of those times, they looked upon them as moral lessons, rather than as scientific evidence. Saint Paul in two different places quotes the saying "A little leaven leaveneth the whole lump." The phrase in analogous to our "One rotten apple spoils the barrel." It seems to be used as an indication that if even a trifling misbehavior or sin enters the soul, that soul will eventually be entirely corrupted—just as one sinner in a group wil eventually corrupt the whole crowd.

In fact, the new soft "leavened bread" may have been viewed as the result of corruption—or perhaps it was just a matter of the strong grip of tradition on religious practices—for we still use flat, unleavened bread on certain occasions, such as Passover, or in communion wafers. (I suspect, though, that modern unleavened bread is better than the prehistoric variety. At least, I eat matzos with pleasure at any time of the year.)

Yeast will also convert fruit juice to wine and soaked barley sprouts to beer, and that, too, is older than history.

It was not till the early nineteenth century that these fermentation processes came to be studied systematically by chemists.

In 1833, a French chemist, Anselme Payen (1795–1871), separated a substance from grain sprouts that was *not* whatever it was that produced beer. The substance did, however, convert starch to sugar more quickly than such a change would take place spontaneously. Payen called the substance "diastase," from a Greek word meaning "to separate" (though I'm not sure why Payen thought that term was appropriate).

This speeding of a chemical reaction was a phenomenon that had been discovered in the course of the previous quarter-century, and the process had been named "catalysis" (see "The Haste-makers," in *Of Time and Space and Other Things,* Doubleday, 1965), but the substances that brought about catalysis had, till then, been inorganic ones such as powdered platinum. In 1811, a catalytic method had been discovered even for the speedy production of sugar from starch that Payen was studying, but the catalyst had, in that earlier case, been dilute mineral acids.

Diastase differed from those catalysts, earlier known, in being an organic material. It deserved, therefore, a special name. Such organic catalysts came to be known as "ferments," thus indicating a relationship to whatever it was that brought about the fermentation processes that produced beer, wine, and leavened bread.


It was known at that time that there was something in the stomach lining that broke up, or "digested," protein molecules. In 1836, the German physiologist Theodor Schwann (1810–1882) treated stomach linings in such a way as to isolate the active principle. It was another ferment and he called it "pepsin," from the Greek word for "to digest." It was the first ferment to be isolated from animal tissue.

Obviously, leaven was (or contained) a ferment, too; one that catalyzed a reaction that converted the starch of grain, or the sugar of fruit juice, to carbon dioxide and alcohol. Still, there was a difference between leaven and ferments such as diastase and pepsin. Diastase and pepsin existed in definite quantity and were eventually used up. Leaven, on the other hand, seemed to form more of itself indefinitely. A little leaven leaveneth the whole lump.

Schwann came up with a notion about this, one he reached rather indirectly.

He began, actually, by considering putrefaction. He noticed that if meat was boiled and then kept in heated air, it did not putrefy. Schwann felt that there were microorganisms that existed in the meat, and in the air, and that these brought about putrefaction. Heat killed the microorganisms and therefore stopped the putrefaction.

There were other scientists, however, who thought that putrefaction was not brought about by microorganisms, but by oxygen, and the heat somehow damaged the oxygen. To test that, Schwann heated air and then let a frog breathe it. The frog got along perfectly well on the heated air so Schwann didn't think anything was wrong with the oxygen.

To test this further, Schwann suspended leaven in water, boiled it, and then supplied it with heated air and expected to see that it would still ferment sugar and starch, thus again proving the oxygen to be undamaged. *But that didn't happen. The fermentation was stopped.*

Schwann had to reverse his attitude. It was known that leaven contained microscopic little globs that just sat there and did nothing, so that no one thought of them as living. However, since heat apparently stopped the leaven from working, Schwann announced, in 1837, that the globs must be living yeast cells that could be killed by heat.

He was just beaten to the punch here by a French physicist, Charles Cagniard de la Tour (1777–1859), who, looking at the globs in leaven under the microscope, had actually seen them bud and produce new cells. They lived and reproduced, so no wonder a little leaven leaveneth the whole lump.

This view met with resistance, however, from the foremost chemists of the day. The German chemist Justus von Liebig (1803–1873) was a particular opponent. He insisted forcefully that fermentation was a chemical process and not a biological one, and so great was his prestige that he carried his point for twenty years.

But then came an even greater than Liebig—the French chemist Louis Pasteur (1822–1895). He investigated fermentation in great detail, studying leaven carefully under the microscope and indulging in subtle experimentation. He found, for instance, that leaven could not carry out fermentation if its environment was lacking in nitrogen, something that was to be expected of living material. By 1857, he had shown, beyond any question or doubt, that leaven, in carrying through fermentation, absorbed nutrients, grew, and reproduced—in short, that it consisted of living yeast cells.

In 1875, a German biochemist, Wilhelm Friedrich Kühne (1837–1900), isolated another digestive ferment. This one was from pancreatic juice, and Kühne called it "trypsin," from another Greek word that implied "digestion." Like pepsin, trypsin broke down protein molecules, but the two ferments were not identical, for whereas pepsin worked only in strongly acid solutions, trypsin worked only in slightly alkaline solutions.

In the light of Pasteur's work, Kühne decided there had to be two kinds of ferments. There was one kind that worked only as part of a living cell such as yeast ("organized ferments") and another kind that could be extracted from tissue and would do its work even though it was not part of anything actually living ("unorganized ferments").

Kühne felt the distinction to be a fundamental one, so that it was worth making in the scientific vocabulary. He suggested, in the same year in which he discovered trypsin, that the word ferment be restricted (for historic reasons) to the substances in living cells. The unorganized ferments, like diastase, pepsin, and prypsin, he suggested, be called "enzymes" from Greek words meaning "in yeast." It was a poor name for his purposes since the unorganized ferments were *not* in yeast. What he meant, though, was that they resembled, in action, those ferments that *were* in yeast. In any case, the word enzyme, now such an integral and well-known part of even ordinary language, thus came into use in 1875.

It was useless to make this distinction, however, unless it *was* a distinction. It was important to show that any destruction of the integrity of the yeast cell would stop fermentation. Heat would do the trick, of course, but it would be still more impressive if simple mechanical destruction of the yeast cell—just pulling it apart into fragments, at room temperature—would stop the fermentation. If that could be shown, then it was reasonable to suppose that the ferment was not simply a substance within the yeast cell, but was the work of the cell as a whole.

Taking up that task, in 1896, was a German chemist, Eduard Buchner (1860–1917), and he did so at the suggestion of his elder brother, Hans, who was himself an eminent chemist.

What Buchner did was to place yeast in a mixture of sand and diatomaceous earth and grind the whole in a mortar vigorously. Undoubtedly, the yeast cells would all be punctured and shredded by this sort of treatment, though individual molecules, it might be surmised, would not be affected.

By this treatment, Buchner quickly converted the leaven into a thick paste. He wrapped the paste in thick canvas and placed the whole under high pressure, so as to squeeze a fluid out of it. This fluid represented the liquid contents of the yeast cell and, when Buchner studied the liquid under the microscope, no intact yeast cells could be found in it.

Buchner was quite convinced, even in advance of testing the matter, that this fluid would have no fermenting effect at all, but he didn't want it to go bad. He wanted no infestation by microorganisms to induce chemical changes and render his results dubious. Nor did he want to be in a situation where he had to spend all his time grinding and squeezing new batches so he could conduct his experiments with fresh fluids only.

One way of preserving a tissue extract against bacterial action is to dump a lot of sugar into it. While bacteria like sugar even as you and I, making a solution stiff with it is going too far for them. (It is this same principle that is used by people who prepare fruit preserves, jams, and jellies. The sugar preservative not only prevents it from spoiling but makes the entire preparation taste heavenly to children—or to those with youthful hearts, such as I.)

Buchner dumped his sugar into the yeast juice, therefore, and I am fond of imagining that he jumped at least five feet after he did so, for the sugar solution started fermenting. It was precisely what he did not expect.

It was clear, then, that yeast contained a ferment that could be withdrawn from the cell, and would still do its work precisely as it had done when it was part of the cell. Buchner named the ferment "zymase," from the Greek word for "yeast."

It could be seen that there was not true distinction between ferments and enzymes. Any ferment that was inside a cell could, with the proper treatment, be extracted from the cell without losing any of its catalytic abilities. Biochemists might as well call them all ferments, or all enzymes—and the decision was to call them all enzymes.

Buchner won the Nobel Prize in chemistry in 1907 for this work. He then went on to volunteer for service the instant World War I broke out, even though he was fifty-four at the time. The German authorities were stupid enough to accept his quixotic gesture and he was shot dead in 1917 on the Rumanian front. Surely the Germans could have found greater use for his brains than to place him in the front lines as a bullet-stopping device. (Nearly half a century earlier, when Pasteur, at forty-eight, had tried to volunteer for service in the Franco-Prussian War, the French patted him soothingly on his head and told him he was worth more to the nation and the world in his laboratory.)

Now that enzymes could be defined, without reference to living cells, as "organic catalysts," the question was, What were they?

There are a vast number of different types of organic compounds. Were enzymes spread all over the lot, or were they members of a close-knit group of one type or another?

This was not something that was easy to determine. Catalysts, in general, work in a very small concentration, and even that very little goes a long, long way. Catalysts don't necessarily take part in a reaction, but sometimes merely offer a surface upon which a chemical reaction can, for one reason or another, take place easily. As I like to describe it, they are like a small desk top on which you can rest a paper sheet and much more easily write a note than you could upon the same paper sheet suspended in midair. You need but one such desk top on which to write a million notes, and amid the blizzard of paper it might well be difficult to locate that desk top catalyst.

What most chemists assumed, though, was that enzymes were proteins. The proteins, of all the different types of organic material, possessed the most complicated molecules. It was easy to visualize each one as having a molecular surface of a distinct and characteristic shape. A proper surface would just fit certain reacting substances and allow them to react much more quickly than they would otherwise. What's more, protein molecules could have such precisely shaped surfaces that each could fit only one molecule, and no other. That would account for why an enzyme might be able to catalyze a reaction involving one particular molecule and no other. That is what is called the "specificity" of enzyme action.

The enzyme theory seemed to be a perfect explanation for enzymes, but it did have the drawback that no one seemed to be able to prove it. In fact, the most careful investigation seemed to end in disproof.

The German chemist Richard Willstätter (1872–1942) undertook to investigate the matter between 1918 and 1925. He purified a solution of a variety of different enzymes. In each case, he got rid of inert impurities, leaving himself with a solution of undiminished enzyme strength, but with less and less dissolved material in it as the purification process went on.

In the end, Willstätter found himself with clear solutions that exhibited full enzyme activity but that showed no signs of protein content. Using the most delicate tests for protein in his repertoire, Willstätter came up with a definite negative. He concluded that enzymes were not protein in nature, but were probably small molecules. In view of many of the properties of enzyme action, this seemed a dubious conclusion, but Willstätter was a crackerjack chemist who had won the Nobel Prize in chemistry in 1915 for his work on chlorophyll and other plant pigments, and few like to argue the point with him.

And yet even as Willstätter was working toward his conclusion, an American biochemist, James Batcheller Sumner (1887–1955), was working toward another.

Sumner was particularly interested in an enzyme that broke down the waste product urea into the still smaller molecules of ammonia and carbon dioxide. The enzyme was called "urease." (The *ase* ending, first used by Payen in diastase, has become universal for the names of enzymes and enzyme groups, with a very few exceptions in the case of those enzymes that,

163

like pepsin and trypsin, became well known before the convention was firmly established.)

A seed called the "jackbean" was particularly rich in urease, and while Willstätter was purifying his enzyme solutions, Sumner was purifying his jackbean extract. It took Sumner nine years to learn to purify it in a satisfactory manner, but at the end of that period he obtained small crystals, which, on solution, showed very strong urease activity.

Sumner decided that those crystals were actually crystals of urease—the thing itself. When he tested those crystals, they reacted strongly positive to tests for proteins. His conclusion, in 1926, despite Willstätter's work, was that urease was a protein. What's more, if one enzyme is a protein, it seems a reasonable possiblility that others are too; and it is even possible that all of them are.

Willstätter shook his head. He dismissed Sumner's work rather cavalierly. Willstätter was famous and highly regarded and Sumner was a relative nobody. Sumner's work, therefore, was not accepted for several years.

Also interested in the matter, however, was another American biochemist, John Howard Northrop (1891–    ). Following the line of Sumner's work, he crystallized pepsin in 1930. Then, in 1932, he crystallized trypsin; and in 1935, chymotrypsin (still another digestive enzyme). All were found to be proteins.

What's more, Northrop's precedures were simple and systematic and could be easily followed by anyone. A large number of enzymes have been crystallized since and all have proved to be proteins.

The matter was settled beyond all doubt, and Willstätter was wrong. In 1946, Sumner and Northrop won shares of the Nobel Prize for chemistry.

What was wrong with Willstätter? He was a first-class chemist and would not make foolish mistakes. Actually, he didn't. He ended up with an enzyme solution showing strong activity, and very little in the way of impurities. That solution, however, contained so few enzyme molecules (very few are needed, after all) that even the most sensitive protein test at Willstätter's disposal had really failed to indicate anything. His work was meticulous and his conclusions reasonable—but his is an example of the unreliability of a negative result. Showing something to be not-A is never comfortably sufficient unless you also show it to be B.

Sumner and Northrop, on the other hand, managed to treat the solution in such a way as to get the enzyme out in solid, crystalline form. They then dissolved it in the smallest convenient amount of water and in that way obtained a concentrated solution that gave a positive reaction to all tests for proteins. That is all very easy to see—after the fact.


Proteins, as it happens, are made up of chains of amino acids. A number of them are made up of nothing else than that; and they are "simple proteins." Among the enzymes, pepsin and trypsin are examples of simple proteins.

Some proteins, however, are made up of amino acid chains, plus portions that are *not* amino acid chains. They are called "conjugated proteins." Some enzymes are conjugated proteins. Examples that I have not previously mentioned in this essay are "catalase," "peroxidase," and "cytochrome oxidase."

If the non-amino acid portion is firmly attached to the protein, it is called a "prosthetic group." In some enzymes, however, the non-amino acid portion is not firmly attached to the protein but is easily removed. The removed portion is called a "coenzyme," and it is the coenzyme that is highly significant in connection with vitamins (aha!).

We will consider the connection of coenzymes and vitamins in the next chapter.

# 10.

# THE BIOCHEMICAL KNIFE-BLADE

I was in a theater not too long ago, waiting for the curtain to go up, and a white-haired woman approached me and said, "Dr. Asimov, we were schoolmates once."

"Really?" I said, with my usual suavity. "You scarely look old enough."

"But I was," she said. "In P.S. 202."

I was galvanized, for I had been at P.S. 202 between the ages of eight and ten. I told her that.

"I know," she said. "I remember you, because the teacher told us once that a certain city was the capital of a certain state, and you piped up at once and told her she was wrong, and that a different city was the capital. She argued with you and, at lunch time, you dashed home and came back with a big atlas and proved you were right. I never forgot that. Do you remember it?"

To which I replied, ruefully, "No, in all honesty, I don't, but I know I was that boy just the same, for I was the only kid in school stupid enough to offend and humiliate a teacher just because I refused to pretend I was wrong when I knew I was right."

Then, at the intermission of that play, I proved that I was still as stupid as ever. A second woman approached me and asked for my autograph on the playbill, and I acceded, of course.

She said, "Yours is only the second autograph I have ever asked for, Dr. Asimov."

"Whose was the other?" I asked.

"Laurence Olivier's," she said.

I smiled and opened my mouth to thank her, but I heard myself say, "How honored Olivier would feel if he knew the company he kept."

It was intended as humor, of course, but the woman walked away silently and without as much as the tiniest smile, and I knew I had just further reinforced my reputation for monstous vanity.

Don't think, then, that I don't feel a distinct twinge every time I sit down to write one of these chapters, wondering, as I do, whether my natural stupidity will show up too clearly this time. Let's hope it doesn't as I write the fourth and last of my vitamin-related chapters.

A protein molecule is made up entirely, or almost entirely, of one or more chains of "amino acids."

At one end of an amino acid is an "amine group," made up of a nitrogen atom and two hydrogen atoms ($-NH_2$). At the other end is a "carboxylic acid group" made up of a carbon atom, two oxygen atoms, and a hydrogen atom ($-COOH$). (That's why it's called an amino acid.)

In between the amine group and the carboxylic acid group is a single carbon atom that is bonded to each. That carbon atom has two additional bonds, one of which is attached to a hydrogen atom, and the other to a "side-chain."

This side-chain can be another hydrogen atom, or it can be one of a variety of carbon-containing groups of atoms. The various amino acids that are found in protein molecules differ from each other in the nature of their side-chains. There are twenty different amino acids that are to be found in almost any protein molecule you isolate from living tissue, and each has a different side-chain.

Amino acids tie together when the amine group of one combines with the carboxylic acid group of another. A long succession of such hookups makes a chain of amino acids, and the important thing about such a chain is that the side-chains remain untouched and stick out of the chain like charms on a bracelet.

Every amino acid chain has a natural tendency to curve, bend, and double up at particular places, thus forming a three-dimensional object with the side-chains sticking out here and there like fuzz. Some of the side-chains are small, some are bulky; some have no electric charge, some have a positive electric charge, some have a negative electric charge; some have a tendency to dissolve in water but not in fat, some have a tendency to dissolve in fat but not in water.

Each different arrangement of amino acids produces a protein with a different pattern of side-chains on the surface; and each different pattern of side-chains signifies a protein molecule of distinctively different properties.

The number of possible arrangements in a chain made up of hundreds of different amino acids of twenty different varieties is unimaginable. If the chain contained only twenty amino acids, one of each type, the number of arrangements would be a little over 2,400,000,000,000,000,000 (two and a half billion billion).

Imagine the number of different arrangements possible if there were dozens of each kind of amino acid scattered randomly along the chain. I once calculated that the amino acids in a molecule of hemoglobin could be arranged in any of $10^{620}$ ways. (That's a one followed by 620 zeros.) The number of all the hemoglobin molecules that have existed in all the hemoglobin-containing organisms that have ever lived on the earth throughout its history is nothing compared to that number. Even the number of all the subatomic particles in the universe is as nothing compared to that number.

It is not surprising, then, that protein molecules can produce a virtually endless number of surfaces, so that it is relatively easy to find one that is well suited to any particular function. That is what makes the chemistry of life so versatile and delicate a thing, and why, starting with the simplest of protein molecules over 3 billion years ago, life could vary itself into some tens of millions of different species, at least 2 million being now alive.

Some particular proteins are very common and make up a huge mass of material in living things generally, There is, for instance, the keratin found in skin, hair, horns, hooves, and feathers; the collagen found in cartilage and connective tissue; the myosin found in muscles; and the hemoglobin found in blood.

If we disregard sheer bulk, however, and simply consider all the different kinds of proteins known, by far the great majority of them are enzymes. There are about two thousand different enzymes that are known and have been studied, and very likely many more that biochemists have not yet isolated. What's more, each enzyme may exist in a number of slightly different varieties.

Each enzyme has a surface that presents a characteristic shape, electric charge pattern, and chemical tendency. Each one, then, is able to attach itself to only one of a very few closely related molecules, or even to only one altogether, and to supply the environment necessary to make a rapid chemical change possible for those few, or that one, only. In the absence of that enzyme that same chemical change could still take place, but only very slowly.

Since the number of such surfaces actually known to exist and be useful are as nothing to the number that can potentially exist, there is ample room for further evolution and for the formation of endless new species.

Even if millions of planets in our galaxy are riddled with life based on protein molecules, you can see that each planet might have millions of species totally different from those on any other. In fact, the possibility of duplication to the point of familiarity (let alone interbreeding) is just about zero.

* * *

The side-chain pattern is enough to make it possible for a protein molecule to do its job very efficiently, and some enzymes consist of nothing but amino acid chains. The digestive enzymes, pepsin and trypsin, which I mentioned in the previous chapter, are of this type. Such proteins, made up of amino acids and nothing more, are "simple proteins."

It is possible, however, for a protein to include atom groupings that are *not* amino acids within their molecules. Usually, the preponderance of the molecule is indeed made up of amino acids, so that we still think of it as a protein, but the non-amino acid portion can nevertheless be important, even crucial, to its functioning.

Enzymes containing groupings that are not amino acids are "conjugated proteins." (Conjugated, from a Latin word meaning "joined together," is an apt term since the non-amino acid grouping is joined together with the amino acid chain.)

There are various types of conjugated proteins, differentiated among themselves by the nature of the non-amino acid grouping. Thus protein molecules joined with nucleic acids are "nucleoproteins"; those joined with fatlike compounds are "lipoporteins"; those joined with sugarlike compounds are "glycoproteins"; those joined with phosphate groups are "phosphoproteins"; and so on.

The non-amino acid portion of a protein may be attached rather strongly to the amino acid chain, and the attached portion is then known as a "prosthetic group." (Prosthetic is from a Greek word meaning "added to." The prosthetic group is added to the protein molecules, you see.)

Sometimes, however, the prosthetic group is but loosely attached to the protein molecules and can be removed by even gentle treatment. This if often true in the case of enzymes, and the easily detached prosthetic group is then called a "coenzyme," for reasons I shall explain shortly.

Even when an enzyme possesses a coenzyme with a structure that is worlds different from that of proteins, it is still the amino acid chain of the enzyme that supplies the necessary surface and determines enzyme specificity (the ability of an enzyme to work with but a single kind of molecule, or, at most, with a very small number). With the proper molecule singled out, the coenzyme can then do the actual work of bringing about the desired chemical change.

As an analogy, you might consider the enzyme a wooden club, which can in itself, with no addition, do a job well—like bashing an enemy over the head to make him see reason. On the other hand, you can stud the head of the club with nonwooden objects—like bone, or stone, or metal—and these will serve to make the bash a more authoritative one. Or you can attach a sharp blade to a wooden club in such a way as to make a knife or an axe out of it.

The handle isn't very useful in itself when it comes to performing a knife's function; and a knife-blade all by itself would be difficult to manipulate. The two together, however, do the job marvelously well.

Viewed in that way, the amino acid portion of an enzyme is the knife-handle, while the coenzyme is the knife-blade, the cutting edge— But, remember, some enzymes (like some clubs) don't need added material to do the job.

In studying enzymes, it is usually desirable to get one as pure as possible. This is not an easy task. A given enzyme exists in very small concentrations in the cells. Present with it are many other enzymes, together with proteins that are not enzymes, to say nothing of other large molecules such as nucleic acids, and small molecules such as those of sugars, fats, individual amino acids, and so on.

A variety of ways of separating proteins from each other and from other large molecules have been developed, and, by picking and choosing among the fractions to see which one beset brings about the reaction in which you are interested, you can gradually select the enzyme you are after, and obtain it in relatively pure and concentrated form.

However, you also want to get out all the small molecules. You want the molecules of enzyme, and nothing else except the water that keeps it dissolved. (Ideally, you don't even want the water, but would be pleased to obtain the enzyme molecules in crystalline form—just enzyme and nothing else at all.)

To get rid of the small molecules, biochemists make use of "semipermeable membranes." These are thin membranes of the type now used to make sausage casings. They are so thin, and their molecules pack together so loosely, that tiny holes are present. These holes are invisible, of course, for they are of molecular dimensions. They are too small, in fact, to allow a large molecule like that of a protein (made up of hundreds, or even thousands, of atoms) to pass through, but small molecules made up of no more than a few dozen atoms *can* get through. That is why the membrane is said to be semipermeable; it is permeable to some molecules, but not to others.

Suppose, then, that a quantity of enzyme solution is put into a bag of semipermeable membrane, which is then tied off. The bag is suspended in a large beaker of water. Some of the small molecules inside the bag manage, simply by random motion, to find their way through the holes of the membrane out into the water. More and more of the small molecules do, while the large enzyme molecules stay put.

Of course, it is also possible for the small molecules, once in the outside water, to drift back through the holes into the bag of enzyme. Eventually, an equilibrium is set up, with the small molecules moving in both directions at equal rates, so that there is no further change in concentration. However, since the volume inside the bag is usually considerably smaller than the volume outside the bag, most of the small molecules are outside in the water by the time equilibrium is reached.

If not enough of the small molecules have been removed at equilibrium, you can always place the bag of enzyme solution into a new sample of water and set up a new equilibrium that will bring the concentration of small molecules inside the bag to a still lower level. In fact, you might even keep water running into the beaker at one end and out the other so that the bag of enzyme solution is always in the presence of new water. Then, virtually all the small molecules are removed.

This process is called "dialysis" (from Greek words meaning "to loosen through") because you can view the small molecules as being loosened from their association with the large ones and passed through the membrane.

In 1904, an English biochemist, Arthur Harden (1865–1940), was busily purifying the enzyme zymase (which I mentioned in the previous chapter). He used dialysis as one of his methods. He placed a solution of zymase in a bag of semipermeable membrane, and placed the bag in a beaker of water. In that way, he got out most of the small molecules.

When he did this, however, he discovered, to his astonishment, that the zymase inside the bag did not bring about fermentation any longer. However, if he added the water outside the bag to the zymase solution, the mixture was once again active.

Apparently, the enzyme consisted of two parts that were so loosely bound together that even the gentle action of dialysis was sufficient to separate them. One part was made up of large molecules that could not pass through the membrane, while the other part was made up of small molecules that could, and both together were essential to the process that brought about fermentation.

Furthermore, the zymase inside the bag could be made inactive by heat, indicating it to be a protein. A protein molecule is so large and complex that it is rather rickety, so to speak. The vibration of its different parts, made more intense as the temperature rises, soon upsets its organization, disrupts the molecular surface, and naturally destroys the enzyme activity. Cooling does not, of itself, restore the activity of zymase so inactivated, nor does the addition of the material from the water outside.

The material outside the bag can be brought to a boil, however, and, after it cools to room temperature again, it is still capable of activating the zymase (provided the zymase has not itself been heated). The outside material, then, is not a protein.

The enzyme, Harden concluded, is made up of a protein portion and a nonprotein portion. The nonprotein portion he called "cozymase," a word in which the prefix *co-* is from Latin, meaning "together," since the small prtion works together with the large.

For this finding, and for his other work on fermentation, Harden received a share of the Nobel Prize in chemistry in 1929.

The cooperative working of two parts, a large protein and a small nonprotein, was eventually found to be characteristic of a number of enzymes (but not all). In the case of those enzymes made up of two such parts, the protein portion was called an "apoenzyme," the prefix *apo-*, from the Greek, implying "off" or "separation." It is the part of the enzyme that remains when the smaller portion is taken off. The nonprotein portion is called a "coenzyme," and Harden's cozymase came to be called "Coenzyme I." The two portions together make up the "holoenzyme"; the prefix *holo-* is from the Greek and means "complete" or "entire." Actually, apoenzyme and holoenzyme are rarely used, but coenzyme has become a familiar word in biochemistry.

Sharing the 1929 Nobel Prize with Harden was the German-Swedish chemist Hans Karl von Euler-Chelpin (1873–1964), who also did noteworthy work on fermentation. Euler-Chelpin went on to tackle the problem of the molecular structure of Coenzyme I. He began by isolating Coenzyme I from yeast, purifying it, and concentrating it over 400-fold. Finally, he had enough for a detailed analysis, which he completed in 1933.

It turned out that Coenzyme I had a strong resemblance to the nucleotide structures that occurred in nucleic acids, but differed from them most notably in containing as part of the structure a pyridine group made up of a ring of five carbon atoms and one nitrogen atom. It also contained two phosphate groups, so that it could be called "diphosphopyridine nucleotide," usually abbreviated DPN.

Another coenzyme, one that was called Coenzyme II, differed from DPN only in the presence of a third phosphate group, so that it was called "triphosphopyridine nucleotide" or TPN.

There are some two hundred known enzymes that have DPN or TPN as coenzymes. DPN and TPN act to remove a pair of hydrogen atoms from one molecule and transfer them to another. This type of chemical reaction is vital to energy production, and the enzymes bringing it about are called "dehydrogenases."

The protein portion of a dehydrogenase provides the surface on which some one particular molecule finds itself at home. The two hundred different apoenzymes make it possible to deal with two hundred different molecules, and on each one of these the DPN or TPN coenzyme will do the work of transferring hydrogen atoms. The DPN or TPN is the biochemical knife-blade that does the "cutting," then, but that requires the selective apoenzyme "handle" to be a useful tool.

The most interesting thing about DPN and TPN is that the pyridine ring forming part of the molecule, when separated from the rest, proves to be a molecule of nicotinamide, which, as I mentioned in chapter 7, is the vitamin whose absence from the diet produces the deficiency disease pellagra.

If nicotinamide is missing from the diet, the body cannot form DPN or TPN, and that means the dehydrogenases begin to halt in their functioning, and the cells fail to function normally. The symptoms of pellagra are merely a series of signs of this failure.

What's more, as biochemists determined the structure of more and more coenzymes, it turned out that various vitamin were often included in their structures. A vitamin is needed in the diet, then, in order to form a coenzyme that will allow some key enzyme or enzymes to work. Without the vitamin, some key reactions will fail within the cells, so that disease and, eventually, death will result

Since enzymes are catalysts, they are needed in the body in only small quantities. Coenzymes are therefore also needed in only small quantities, and, in consequence, vitamins are needed in only small quantities—but those quantities, however small, are vital to life just the same.

(Some enzymes work properly only in the presence of a metal atom, and that is the reason for the essentiality of trace quantities of certain metals, such as copper, manganese, and molybdenum, in the diet. Again, there are poisons that act very quickly and in small doses to end human life. They work by combining with key enzymes or coenzymes in such a way as to prevent their functioning.)

But why is it that the human body cannot form the nicotinamide portion of Coenzyme I? It can, after all, form the rest of the molcule without trouble.

Some forms of life can form, without exception, *all* the complex molecular structures needed for their functioning, using as their starting materials very simple molecules present in the environment even before any life existed at all.

Plant cells, for instance, start with water, carbon dioxide, and certain mineral substances present in the sea and in the soil, and make use of the energy of sunlight, also present from the beginning. From such a start they manufacture all the substances they need.

Microorganisms and animal cells that cannot use sunlight as the ultimate energy source must obtain energy by oxidizing organic materials that were originally produced by plants. Given this energy, they can start with relatively simple materials and build up the complicated molecules they need. Nevertheless, you can see that they depend on the plant world for energy and, therefore, for life.

(Some few microorganisms are "chemosynthetic" and can obtain energy by taking advantage of chemical reactions that do not involve organic substances.)

Suppose that a particular molecule is needed by an organism in small quantities and can be absorbed as such from the food it eats. The organism might lose the ability to make the molecule and come to depend on dietary supplies. The more advanced and complex an animal, the more likely it is to do this.

Why should that be? My own feeling is that the

more complex an organism, the more enzymes are needed to make everything possible. For instance, animals have muscles and nerves, which plants don't have, and must make use of enzyme-mediated reactions that plants can do without. Room must be made for the various enzymes that control myriads of reactions in complex organisms that simple ones need not deal with.

If, then, there are some cell substances that are required to do only a very small extent, why bother manufacturing them? Let the diet supply that and thus leave room for other, more necessary chemical processes. (In fact, you might argue that animal cells, by doing without the complex machinery required for photosynthesis, and eating plant cells instead so as to get energy from their diet rather than from the sun, make room for the more complex animal functions.)

Naturally, some things can't be skimped on. If a particular small molecule is needed in quantity, the diet cannot be sufficiently depended on to supply those quantities. It would be too great a risk to take. It is only when small quantities are needed that the risk is reasonable.

Thus, of the twenty amino acids found in proteins generally, the human body can build up twelve from fragments of other molecules that it finds in its food. If the diet is short in one or another of those twelve, the body can make that up from its own resources, a* the expense of maintaining the battery of enzymes that make it possible.

The remaining eight of the amino acids cannot be formed by the human body, however, and must be found in sufficient quantity in the diet. These eight are therefore known as the "essential amino acids," not because they are more essential than the others to the body's workings, but because they are essential components of the diet, if deficiency disease, and death, is to be avoided.

Why those eight? Because they are the eight needed in least quantity, so that it is safer to gamble on those than on the rest.

As it happens, the vitamins contain atom combinations that do not occur elsewhere in the body. The body makes use of a nicotinamide grouping of atoms only in Coenzymes I and II and nowhere else. Why maintain enzymes for the synthesis of such a grouping? Get it instead by eating some less complex organism that, for some reason, must maintain the necessary enzyme investment.

How does the body know what substances it can safely gamble on finding in the diet and what substances it cannot? It doesn't.

Every once in a while, an organism is born without a particular enzyme or other, as a result of a random mutation. If that enzyme lack deprives it of the ability to make substances it cannot depend on its diet to supply inadequate quantities, it quickly dies. If the enzyme that is lacking happens to control the formation of something needed only in traces, however, the organism may get it from its diet and will then continue to live. It may even benefit as other chemical abilities find room to flourish.

Naturally, this purchase of more efficient complexities is at the price of having to be more careful with the diet than we would otherwise have to be, but apparently the benefit is cheap at the price. Most animals, with their diet restricted to what they can find in nature, are guided by their instincts and their taste buds to eat that which will supply them with what they need.

Human beings, on the other hand, have the ability to fool around with their food, refining items to keep those parts that taste best or keep best, and dumping the rest. They indulge in boiling, frying, roasting, salting, drying, sugaring, and other things that make food taste better or keep longer—and, in recent years, the addition of myriads of chemicals. This all tends to make it riskier to depend on diet to supply the substances we can't make ourselves and must have.

On the other hand, we now have synthetic vitamins, mineral pills, and so on. We may still die of deficiency diseases out of the perversity of our tastes, or out of the sheer insufficiency of the quantity and variety of food that our surroundings or our economic status will allow us. But at least we know enough now to avoid such a fate if we are both fortunate and rational.

# PART III
# GEOCHEMISTRY

# 11.
# FAR, FAR BELOW

Some years ago, a Hollywood producer suggested that I write a "treatment" that could be turned into a screenplay, concerning a trip to the center of the earth.

I pointed out that a successful picture had been made on that theme, starring James Mason and Pat Boone. The producer knew about that and pointed out in his turn that the art of special effects had advanced enormously since than, and that a much more spectacular version could now be made.

"One that is scientifically accurate?" I asked.

"Of course," he replied genially, not really knowing what he was letting himself in for.

So I told him. "In that case," I said, "there can be no journey down long caves; no hollows deep in the earth; no inner worlds; no underground seas; no dinosaurs; no cavemen. Earth will be pictured as full of matter all the way down, and with temperatures rising into the thousands of degree."

He wavered and said doubtfully, "Could you make an interesting story out of that?"

"Sure," I said, with the calm confidence of long experience.

"All right," he said.

So I whomped up a treatment that I thought was very interesting and quite scientific, except for the fact that I invented vessels that could bore through solid rock without difficulty and that could remain cool when surrounded by molten iron. (There has to be *some* poetic license.)

I had to fight off attempts to introduce additional nonsense and just as I was beginning to think there would be an honest picture involving the center of the earth, the powers that be in Hollywood turned it down with a shudder I could feel in Manhattan.

I suppose that if another voyage to the center of the earth is done, it will involve a hollow earth; a small, radioactive sun at the center; underground seas; dinosaurs; cavemen; and beautiful actresses in skimpy costumes.

But not with my help!


What made people think that the earth was hollow, to begin with?

The initial spur may have lain in the existence of caves, some of which are quite large and intricate and were not fully explored. Since the explored portion reached considerable depths, it was easy to suppose caves reached even greater depths in places beyond where anyone had the nerve to explore.

Then, too, the common notion of an underworld, in which the spirits of the dead existed, must have given rise to the notion of a hollow earth, once our planet was accepted as a spherical body. Dante's *Divine Comedy* is the greatest literary expression of a hollow earth with hell located in the hollow.

Finally, a hollow earth is a dramatic conception. It makes for interesting stories and gives scope for exciting adventures.

Perhaps the first notable hollow-earth story was that of a Danish writer, Ludvig Holberg (1684–1754), who wrote a story, in Latin, called *Nicholas Klim Underground*. It was published in 1741 and was quickly translated into the various European languages. It placed a little sun at earth's center and described several miniature planets circling, forming a micro solar system.

This notion was translated into "science" by one John Cleve Symmes (1742–1814), who maintained that the earth was not a sphere, but a doughnut. There were large holes at or near the north pole and the south pole, and, presumably, these communicated with each other.

It was safe for Symmes to make this assertion, since in his lifetime the polar regions of the earth were still impenetrable mysteries, and no one could check the existence or nonexistence of a hole in either place. Naturally, Symmes was found to be very convincing by a great many simple souls, for there seems to be a rule that the more foolish an assertion, the more ardently people will believe it. (We know that very well by observing the contemporary world.)

The idea was grist to the mills of science fiction writers. Edgar Allan Poe (1809–1849), in his *Ms. Found in a Bottle*, published in 1833, describes the plight of a ship caught in a gigantic whirlpool in the polar regions. Presumably, the ocean is pouring constantly into the northern Symmes' hole. (It is to be hoped that the water makes its way back to the surface elsewhere or the oceans would have been drained long ago.)

Jules Verne (1828–1905) steered clear of holes in the bottom of the sea, but in *A Journey to the Center of the Earth,* published in 1864, the starting point is still in the far north—a volcano in Iceland. Verne's explorers find an ocean in the earth's interior, and see such exotica as giant reptiles, mastodons, and cavemen.

The most recent notable examples of hollow-earth stories were those of Edgar Rice Burroughs (1875–1950). Beginning with *At the Earth's Core,* first serialized in 1914, he wrote a series of stories about Pellucidar (the name he gave the inner world).

And yet, as long ago as 1798, it was entirely obvious that earth was *not* hollow, and that Symmes was talking through his hat.



In 1789, the English physicist Henry Cavendish (1731–1810) determined the mass of the earth quite accurately, as roughly 6 sextillion tonnes. The best figure that we now have is 5,976,000,000,000,000,000,000 tonnes (nearly 6 sextillion tonnes). From this, and from earth's known volume, we can at once determine that the average density of earth material is 5,518 kilograms per cubic meter.

The density of earth's surface rocks is, however, something like 2,600 kilograms per cubic meter, while the density of the ocean is a trifle over 1,000 kilograms per cubic meter. If, on top of that, the earth were hollow, there is simply no conceivable way in which it could have the average density and the total mass that it does have.

To account for earth's mass, the earth's interior must not only not be hollow, it must be made up of material that is considerably more dense than the stuff on the surface.

Or look at it this way. Suppose the earth had a mass of 6 sextillion tonnes, and all that mass was (in some fashion) concentrated into a relatively thin shell about a central hollow. The gravitational field associated with that mass would be so intense that the shell would crumple and crush together into a sphere (or an oblate spheroid if the body were rotating about its axis). Nor could any hollow exist, since the gravitational field would totally erase it.

To be sure, there are caves on earth, but these are strictly surface phenomena, and are trivial irregularities, much like the mountains and valleys that only insignificantly roughen the earth's smooth oblate sphericity.


Very well, then, we can ignore the madness of pseudoscientists and the romanticism of science fiction writers and consider the earth as dense and unhollowed all the way through. The next question is: What is the earth's interior composed of?

There's no easy answer to that. There's no way we can directly observe the material of the earth more than a few kilometers below its surface. Even today, we are stymied. We can reach 380,000 kilometers across space and bring back material from the moon's surface, but we have yet to bore as much as 15 kilometers into the earth. To probe down the 6,400 kilometers to earth's center may remain utterly unlikely for a long, long time to come.

We can, however, make intelligent deductions from observations on earth's surface. For instance, we know that the earth's outer crust, which we *can* observe directly, is rocky in nature. The simplest conclusion to which we might come, therefore, is that the earth is rocky all the way through. The farther down we go, the denser the rock becomes, since a greater and greater weight of overlying rock presses down upon the deeper layers, which are more and more compressed (and therefore denser) in consequence.

However, it is possible to study the response of rocks to the forces of compression. Even though we have only very recently been able to reach (momentarily) compressions of the order of magnitude encountered at the earth's core, it has become clear that rock will not compress sufficiently. If the earth were rocky through and through, the densities in the interior would simply not be great enough to account for an overall average of 5,518 kilograms per cubic meter. Clearly, the earth's interior must be composed of some material that is denser than rock under zero pressure, to begin with, and that would remain denser than rock at any higher pressure.

Such a material suggested itself quite early in the game.

In 1600, the English physicist William Gilbert (1540–1603) experimented with a sphere he had shaped out of a magnetic material called "magnetite" or "lodestone" (a naturally occuring form of iron oxide) and observed the behavior of compass needles in its vicinity. The compass needles behaved exactly as they did in response to earth's magnetic field and the obvious conclusion was that the earth was itself a spherical magnet.

Why should it have magnetic properties, however? The rocks of the earth's crust are, by and large, non-magnetic, and the exceptional magnetite makes up a very tiny portion of the whole. Suppose, though, that the earth's interior is solid magnetite. Magnetite has a density, at zero pressure, of about 5,200 kilograms per cubic meter, twice that of the common rocks of the crust, and it would be correspondingly denser than those common rocks under the great pressure of earth's interior. Yet magnetite would still not be dense enough.

Suppose, then, the earth's interior were a solid mass of iron. That, too, could be magnetic, and the density of iron, at zero pressure, is 7,860 kilograms per cubic meter, three times that of the common rocks of the crust. That would be dense enough.

About 1820, scientists accepted the fact that meteorites were bits of solid matter that reached earth from outer space. When they studied such meteorites, it turned out that there were two chief types. There were "stony meteorites" and "iron meteorites." The former consisted chiefly of substances not very far removed from the materials making up earth's crust. The latter consisted almost entirely of a mixture of iron and nickel in proportions of nine to one. (Nickel, like iron, has magnetic properties. The mixture would serve as an internal planetary magnet.)

In the 1800's, it was a popular view that the asteroids were the remnants of a planet that had existed in an orbit between those of Mars and Jupiter and that had, for some reason, exploded. It seemed reasonable to suppose that the outer portion of that planet was rocky in nature, and the interior portion nickel-iron, and that these two parts were the source of the two types of meteorites.

In 1866, a French geologist, Gabriel August Daubrée (1814–1896), suggested that earth, too, might have this as its fundamental structure, a rocky exterior wrapped about a nickel-iron interior.

There was more, however, to the earth's deep interior than a chemical difference. For one thing, it seemed clear that the earth's interior was a place of heat. Volcanic eruptions were unmistakable evidence of that. (It was undoubtedly because of volcanic action that the notion arose of hell as being a place of "fire and brimstone.")

In later times, more subtle evidence of internal heat arose. The vast energies of earthquakes had to be fed by something, and internal heat was the most reasonable source to be suggested. Then, too, many rocks on earth's surface have crystallized in fashions that seem to betoken exposure to great temperatures and pressures, presumably because they were deep underground at one time. Furthermore, as human beings dug their mines more and more deeply and observed the results more and more closely, it became clear that temperatures rose steadily with increasing depth.

But where did the heat come from? One theory of

the origin of the earth would have it that the planets of the Solar System were part of the sun to begin with. The earth was thought, therefore, to be at the temperature of the sun at the start, and to have cooled with the years. The outer crust cooled sufficiently to solidify, but rock is a good heat insulator and the interior lost heat only slowly, therefore, and is still hot to this day. Indeed, some scientists tried to estimate the amount of time it would take for the earth to cool off in this way, and decided that the earth could only have an age of some tens of millions of years.

This notion of a sun-born earth gradually weakened. The mechanical details involved in pulling the planets out of the sun and establishing them at their present distances and in their present orbits proved an unexpectedly intractable problem. Furthermore, by the 1920's, it became quite clear that the sun's interior was enormously hotter than its surface, and that gobbets of sun-stuff would not condense into planets, but would evaporate into space.

A competing theory, originally suggested by the French astronomer Pierre Simon de Laplace (1749– 1827) in 1798, was greatly improved and put into a presently acceptable form in 1944 by the German astronomer Carl Friedrich von Weizsäcker (1912–    ).

The present view, then, is that the sun, together with the planets, were all formed simultaneously by the gradual coming together of smaller bodies. The earth's high internal temperature was therefore the result of the conversion of the kinetic energy of all those bodies into heat.

What's more, in the first decade of the 1900's, it was realized that such elements as uranium and thorium, together with isotopes of such more common elements as potassium and rubidium, underwent radioactive breakdown and gave off heat in so doing. The heat per kilogram per second was very small, but the total planetary supply was sufficient to give off considerable heat, and this heat emission continued with only moderate decline over billions of years.

The earth's interior was therefore not cooling as rapidly as one might expect, and the earth's age proved to be not 25 million years or so, but 4,600 million years—and this was the age of the Solar System as a whole.

Whatever the source of the earth's internal heat, or the rate at which it had cooled to the present value, the question remained as to the state of earth's interior.

The original feeling was that the rise of temperature with depth meant that everything below 80 kilometers had to be molten and fluid, so that the earth was essentially a huge ball of liquid surrounded by a relatively thin solid crust. This was argued against by the Scottish physicist Lord Kelvin (1824–1907), who pointed out that such a thin solid crust would be impossibly weak and would be quickly broken up by the tidal influences of the moon and the sun. As it was, the actual effect of the tides on the earth's solid surface seemed to show that the earth, as a whole, was as rigid as steel.

By 1900, then, it was felt that the high temperatures of earth's interior were neutralized, so to speak, by the high pressures. Although the temperatures were high enough to melt rock and metal at ordinary surface pressure, the increasing pressure with depth kept it all solid, even though the earth's temperature at its very center was as high as 6,000°C. The earth, in other words, had to be (it appeared) solid throughout.

This turned out to create a problem. In 1895, the French chemist Pierre Curie (1859–1906) showed that magnetic substances lost their magnetism if temperatures were raised above a certain level (the "Curie point"). For iron, the Curie point is 760°C, and the temperature of the earth's core is certainly higher than that. It would therefore seem that the core could *not* account for the earth's magnetism. For a while, that remained a puzzle.

By the time the 1800s were coming to an end, scientists were beginning to study earthquakes in detail and, quite unexpectedly, found a new technique with which to study the earth's interior.

The first "seismograph" that could usefully serve to detect the waves of vibrations set up by earthquakes was invented in 1855 by an Italian physicist, Luigi Palmieri (1807–1896). The device was greatly improved in 1880 by the English geologist John Milne (1850–1913), who established a chain of seismographs in Japan and elsewhere. With him, the modern science of seismology began.

When an earthquake took place, the vibrations were detected by different seismographs at different times, depending upon the distance of each from the focal point of the quake. In this way, one could measure the speed with which earthquake waves traveled through the earth's crust.

In 1889, the vibrations of an earthquake in Japan

were detected 64 minutes later in Germany. Had the waves traveled along the curved surface of the earth at the speeds they were known to have, they could not possibly have reached Germany in so short a time. The conclusion was that they had taken a shortcut, passing in a straight line, more or less, through the earth's interior.

In 1902, the Irish geologist Richard Dixon Oldham (1858–1936), studying the waves set up by an earthquake in Guatemala, was able to show that the speed at which the waves traveled through the deeper layers of the earth was slower than that at which they traveled through more shallow layers.

The waves, as they traveled through the earth, would respond to changing speed with depth by taking up a curved path, sometimes even sharply curved, much as light waves curved and were refracted in passing from air into glass or vice versa, or as sound waves curved in passing through layers of air of different density or temperature.

The curved path taken up by the earthquake waves as they passed through the interior of the earth allowed them to reach certain portions of earth's surface but not others. A ''shadow zone'' might be created within which the vibrations set up by earthquakes would not be felt, although the vibrations would be felt both closer to, and farther from, the earthquake center than the shadow zone was.

By studying the nature of the shadow zone and the times it took for the earthquake waves to reach different points on the earth's surface, the German geologist Beno Gutenberg (1889–1960), showed in 1912 that the waves underwent a sudden and pronounced decrease in speed, and a subsequent sharp change in direction, when they penetrated beyond a certain depth. This crucial depth, he determined, was about 2,900 kilometers below the earth's surface.

It was a sharp boundary (the "Gutenberg discontinuity"), so that the earth seemed to be divided into two chief regions. There was a central core, a sphere with a radius of 2,900 kilometers, which was, presumably, nickel-iron in composition. Around it, making up almost all the rest of the earth, was a rocky "mantle." The sudden sharp change in the speed of earthquake waves as they passed from the mantle to the core, or vice versa, was the best evidence yet of a sharp change in chemical makeup between the two regions.

Within the mantle, and within the core, the waves traveled in gently curved paths, which indicated an increasing density with depth. Thus, from a surface density of 2,600 kilograms per cubic meter, the density rises as one probes downward through the mantle until, at a depth of 2,900 kilometers below the surface, it is about 5,700 kilograms per cubic meter. As one moves into the core at that depth, the density rises, suddenly and sharply, to 9,700 kilograms per cubic meter and continues to rise until, at the very center of the earth, it is 13,000 kilograms per cubic meter. Such figures fit the notion of a rocky mantle and a nickel-iron core.

In 1909, meanwhile, a Croatian geologist, Andrija Mohorovičić (1857–1936), was studying an earthquake in the Balkans and detected a fairly sharp change in the speed of the waves at a depth of about 30 kilometers below the surface (the "Mohorovičić discontinuity''). Apparently, the rocky mantle had a thin outermost layer, usually called the "crust."

Both crust and mantle are composed of rocky substances, but the details of the chemical structure are different. The crust is high in aluminum silicate, for instance, while the mantle (judging from earthquake data, and comparing the speeds of the waves through the mantle and through rocks of various composition under laboratory conditions) is high in magnesium silicate.

But the question of the state of the earth's interior, solid or liquid, continued to arise. Even into the 1920's, though, the majority opinion was that it was solid.

Not only was it thought that pressure would keep the core solid even at high temperatures, but the new knowledge of radioactivity contributed to the thought. The radioactive substances, uranium, throium, and so on, were all concentrated in the mantle, perhaps even in the upper layers of the mantle, since compounds of those substances mixed more easily with rock than with nickel-iron. It might be, therefore, that the mantle would be hot, but the core might be comparatively cool, even cool enough to keep the iron below the Curie point, and therefore magnetic.

There are, however, two kinds of earthquake waves. Some are "transverse" and vibrate up and down like light waves, moving at right angles to the direction of propagation of the waves. Those are known as "S waves." Others are "longitudinal" and vibrate in and out like sound waves, in the direction of propagation of the waves. Those are "P waves."

Longitudinal waves, such as the P waves, can travel through any kind of matter: solid, liquid, or gas. Transverse waves, such as S waves, can travel through solids, or along the surface of liquids, but cannot travel through liquids or gasses.

Oldham was the first to note the existence of these two kinds of earthquake waves, and by 1914 it seemed to him that he had never detected S waves passing through the core. He began to suspect, therefore, that the core might be liquid.

Gutenberg, on the other hand, was convinced the core was solid and his prestige was so high that it wasn't till 1925 that geologists generally were convinced that S waves did not pass through the core. Even then, though, they hesitated to conclude that the core was liquid.

In 1926, the English astronomer Harold Jeffreys (1891–    ) calculated the rigidity of the mantle from earthquake-wave data and showed it to be considerably more rigid than the earth as a whole (as calculated from tidal data). That meant that the core had to be less rigid than the earth as a whole, and might, therefore, well be liquid. That finally swung opinion to the other side, and from that time on, the notion of earth's possession of a liquid nickel-iron core was established.

A liquid iron core was certainly above the Curie point, but earth's rotation could set up swirls in it, and these swirls could produce electromagnetic effects that, Curie point or no, would account for earth's magnetic field.

Finally, in 1936, a Danish geologist, Inge Lehmann, noted that the P waves that penetrated the core and that passed quite close to the earth's center, seemed to undergo a sudden small increase in speed. She suggested that there was an "inner core" at the earth's center that was a sphere with a radius of 1,250 kilometers.

How does the inner core differ from the outer core? The general opinion is that whereas the outer core is liquid, pressures at the very center of the earth are great enough to solidify the nickel-iron, so that the inner core is solid.

That is where matters stand now, except that there is some argument as to the precise chemical nature of the core. Some maintain that pure nickel-iron may be *too* dense to account for the overall density of the earth, and that the core must contain a significant amount of oxygen to lower that density. It may be, then, that the core consists of somewhat rusty nickel-iron.

Let me conclude, then, by saying that the solid inner core makes up about 0.8 percent of the earth's volume; the liquid outer core, about 15.4 percent; the rocky mantle, about 82.8 percent; and the rocky crust, about 1.0 percent.

In terms of mass, the dense metallic core (outer and inner) together make up just about one-third of the total mass of the earth, while the rocky outer layers (mantle and crust) together make up the other two-thirds.

# PART IV
## ASTRONOMY

# 12.
# TIME IS OUT OF JOINT

It's a hard life being time-bound, but I come by it honestly. When I was a boy, I had to be downstairs to deliver the papers for my father's candy store and that had to be done *on time,* because the customers had to get their papers before they left for work.

What's more, I had to be in school *on time,* or I would be marked "tardy," and after a while that would be reported to my parents. My mother, being European, and therefore under the peculiar impression that crime ought to be punished, would surely whop me, and with no light hand either.

And then, of course, radio programs began *on the minute* and who wanted to miss them.

So it was a golden day for me when I was given my first wristwatch, and came to be a master of time. I could now see what time it was by a mere glance at my left wrist and that meant I would never be late again. Or, at least if I were, I would *know,* first, that I was going to be late, and, eventually, that I *was* late.

It is now a long, long time since my first watch and I have never been without one since. I don't mean I have always had one somewhere around. I mean I have always had one *on my wrist.* Almost always. I remove it, reluctantly, in order to take a shower, and again when I go to sleep (in which case I always have a clock on the nightstand, with an illuminated dial, so that any time my eyes open I know immediately what time it is).

When I am wearing my watch, I doubt that five minutes ever passes without my casting a quick glance at my wrist, for no purpose other than to know what time it is. I may not need to know the time; that knowledge may serve no conceivable purpose; but that doesn't matter. I must know the time.

In my younger days, I remember that this often faced me with an embarrassing dilemma. There I would be, patting a nice young lady on the head, or pinching her cheek (or whatever it might be that I was doing—it's hard to remember that far back), and then that mad desire to know the time would sweep over me. I knew very well that to glance at my watch would be interpreted by the young woman in only one way—that I was bored, and anxious to be rid of her. This would (for some reason) fill her with rage and the proceedings might well come to an end. I also knew very well that no matter how unobtrusively I looked at my watch, or how cleverly I masked the act ("Is that a scratch on my wrist?"), she would know.

I was at times reduced to the craven expedient of trying to alter the rules of the game at the beginning: "Look, baby doll, I have this nervous twitch that makes me look at my wrist every five minutes. It doesn't mean anything."

"Really?" she would be very likely to answer.

"Well, just put your watch on that chest of drawers and turn the dial away from you."

Let me tell you that just about killed the fun, almost.

In any case, let's talk about time.

In the good old days before everyone had a watch that was accurate enough to tell the minute, if not to the second, people nevertheless managed to get along. There was usually a clock (of indifferent accuracy) in the church spire, the highest point in town, so that everyone might see it. The hours were rung on the church bells so that people might hear the time if they happened to be looking in the wrong direction or if something blocked their view. That is why we have our word "clock" from the French word *cloche*, which means "bell,"

Thus, when Falstaff in *King Henry IV, Part One* brags falsely about having killed the valiant Hotspur at the Battle of Shrewsbury, he says they "fought a long hour by Shrewsbury clock."

People who lived in rural areas didn't have even a town clock to go by. In that case, they used the clocks of heaven. Thus, earlier in the same play, a workman is fretful over the lateness of the night. He says: "An [if] it be not four by the day [4 A.M.], I'll be hanged. Charles' Wain [the Big Dipper] is over the new chimney, and yet our horse not packed."

The stars travel regularly across the sky and, from their position and the season of the year, someone like the workman just mentioned can make a rough estimate of the time.

If you point directly upward you will point to the highest part of the sky, relative to yourself; that is, to the "zenith" (from an Arabic word meaning "overhead"). If an imaginary line is drawn north and south through the zenith, it divides the sky into two equal halves between the rising point of a heavenly object and its setting point. That north-south line through the zenith is called the "meridian," from a Latin word meaning "midday."

The reason for that word is that in passing from east to west, from rising to setting, a heavenly body crosses the meridian halfway on its journey, so that the sun, for instance, crosses it at midday. Heavenly bodies don't necessarily pass through the zenith in crossing the meridian. Generally, they pass north or south of the zenith. The sun and moon, viewed from the north temperate zone, always pass south of the zenith. Nevertheless, a heavenly body, crossing the meridian anywhere, does so halfway in its journey across the sky.

If we were to note the moment when a particular star crossed the meridian on a particular night, and then when it crossed it again on the next night, and then on the next and so on, and do so with a good clock, we would find that the intervals were equal in length to a high degree of accuracy. This is not surprising, since the passage of the stars across the sky is actually a reflection of the rotation of the earth on its axis, and that rotation proceeds at a constant rate.

You might wonder, by the way, why we would take the trouble to measure the intervals between the times of crossing of the meridian, when the meridian is an imaginary line that takes some trouble to set up. Why not measure the intervals between star-rise and star-rise, or between star-set and star-set?

For one thing, the horizon on land is broken and uneven, and therefore hard to observe. Even at sea, where the horizon is smooth, there is usually a haze there, and, even if there wasn't, the atmospheric absorption and refraction of light would confuse the issue. Objects are more easily and accurately observed the higher they are in the sky, and, therefore, most easily and accurately observed when they cross the meridian.

The interval between the crossing of the meridian by a star on one night and then on the next is the "sidereal day." (Sidereal is from the Latin word for "constellation" or "star.") It is the length of a complete rotation of the earth relative to the stars; that is, to the universe generally.

The sidereal day is of interest to astronomers, but not to the population generally. Ordinary people are asleep during the night, and even if they are awake, the positions and movements of the stars are of little interest to them.

People, however, are awake during the day and, during the day, it is impossible not to be aware of the position and movement of the sun. On its changing position depend all sorts of activities and, therefore, the moment at which the sun crosses the meridian is important indeed to everyone.

Of course, one can't really watch the sun cross the sky without going blind, but no one has to. The sun produces shadows that can be observed quite easily and comfortably, and those shadows are a perfect key to the movements of the sun.

Suppose you thrust a pole firmly into the ground. At sunrise, that pole will cast a long shadow westward. As the sun climbs in the sky, the shadow will shorten and shorten and (if you are in the north temperate zone) will swing around northward. The shadow passes the pole to the north, being relatively short then, and then begins to stretch out longer and longer eastward until sunset.

Suppose you mark out the shadows of the pole at sunrise and at sunset as two furrows in the ground, and bisect the angle they form, dividing it in two. This is not difficult to do. In that case, it will be found that the line that serves as the bisector will extend exactly north and south. When the shadow falls on that line, the sun is crossing the meridian and it is exactly midday.

Such a pole is called a "gnomon" (the *g* is silent), from a Greek word meaning "know," since it gives us knowledge concerning the time of day.


The ancients learned to set up a gnomon in a bowl set on a pedestal. The gnomon was placed at an appropriate angle toward the north so that the shadow fell on the rim of the bowl and traveled along that rim from west to east. The distance between the sunrise shadow and the sunset shadow was marked off into twelve parts, marking twelve equal divisions of the day, and thus, you had a sundial.

Why twelve? This was a fashion begun by the Sumerians perhaps as early 3000 B.C. They had not yet worked out a good system of dealing with fractions, so they preferred to use numbers that were least likely to leave fractions when broken into smaller parts. Twelve can be evenly divided by 2, 3, 4, and 6, and was useful therefore.

Each division is called an "hour" (from a Greek word meaning "time of day").

Originally, sunrise marked the zero point of measuring the hours, so that the "first hour" was one hour after sunrise, the "second hour" was two hours after sunrise, and so on. Hence, when the Bible speaks of the "eleventh hour," it doen't mean either 11 A.M. or 11 P.M., but eleven hours after sunrise; that is, the last hour of daylight before the "twelfth hour," which was sunset.

The word "noon" is a distortion of the Greek word for "nine" and meant the "ninth hour," which began when the daylight was three-fourths done. It meant, in other words, midafternoon. Perhaps it was associated with mealtime, and when the chief meal of the day shifted to midday, the association with food was stronger than the association with nine, so that noon became midday, or the sixth hour. For that reason we now speak of "forenoon" and "afternoon," or, if we like to speak Latin, we can say "antemeridian" (A.M.) or "postmeridian" (P.M.).

Naturally, since the daylight was divided into twelve hours, the night was, too.

As we all know, for half the year the days grow longer and the nights shorter, while for the other half the days grow shorter and the nights longer. This is true everywhere but at the equator, and the farther one travels from the equator in either direction, the more marked these changes are.

Where the sundial is the method of marking the

hours, the individual hours grow longer by day and shorter by night, or vice versa, depending on the time of year.

However, sundials were not the only time-keeping mechanisms. They had their shortcomings. They didn't work on cloudy days, and while this did not matter in the almost cloudless climate of Egypt, where the sundial may have been invented, it was a drawback in more atmospherically turbulent regions. Then, too, even in Egypt, sundials didn't work at night.

So people sought other ways of telling time. They considered processes that continued at a slow but apparently constant rate and tried to synchronize them with the sundial.

For instance, candles of a given height and thickness could be manufactured and allowed to burn. Marks could then be made on a second candle, that was not burning, scoring the places reached by the burning candle at the end of successive hours. Similar candles are all marked in this way and from then on the hours can be followed at night by burning candles. In the same way, periods of time can be marked by the sifting of sand or the dripping of water through small apertures.

Such portable devices, if they must measure hours that grow longer and shorter with the seasons, become impractically complicated. It proved much simpler to consider the hours to be of constant length through the day and night, as well as throughout the year. Each hour was a length of time equal to $\frac{1}{24}$th of a day, and this practice has continued to the present.

* * *

There is a question of when the day starts. It seems very natural to begin the day at sunrise; or else, to end the day at sunset and begin a new day at that time.

The peoples of southwestern Asia, including the Jews, began the day at sunset, and this habit continues in the Jewish religious calendar right down to the present. Thus, the Jewish Sabbath (usually considered to fall on Saturday) actually begins at sunset on Friday.

There is even a fossil remnant of that view in Christian life. We speak of Halloween (All Hallow's Day evening), Christmas Eve, and New Year's Eve. These were not originally the evening *before* the holiday itself. They were the first part of the holiday itself that began, originally, at sunset of the previous evening.

For astronomers, however, the imperfections of considering the intervals between sunrises, or between sunsets, were irritating. The moment of sunrise, or sunset, varied with the nature of the horizon. It took the sun a little extra time to rise above a hill on the eastern horizon, and it set a little sooner behind a hill on the western horizon. Besides, clouds and haze often obscured the horizon at the crucial moment. Then, too, as the days grew shorter, sunrise took place a little later each morning and sunset a little sooner, while when the days grew longer, the reverse was true. In either case, the interval from sunrise to sunrise, or from sunset to sunset, was longer at some times than others.

The exact time of the meridian passage of the sun is a lot easier to measure than that of either sunrise or sunset. What's more, the interval between is constant the year round, for as days shorten or lengthen, they shorten or lengthen at both ends equally, leaving the middle in place.

Therefore, the time interval marking the "solar day" (one complete rotation of the earth relative to the sun) is best measured from noon to noon or from midnight to midnight. The choice fell upon midnight because that meant the day changed from one to the next when everyone was sleeping (or should have been) and not in the middle of the active day, something that would upset business records and make them more complicated.

It would make sense, then, to count the hours from one to twenty-four, and this is done under some conditions and in some places. However, the old, old habit of two periods of twelve hours each is too firm to shake off altogether. We usually count, therefore, from 1 A.M. to 12 noon, and then start over and count from 1 P.M. to 12 midnight.

In this way, we no longer count twelve hours of daylight and twelve hours of night. Instead, both ranges of twelve hours are partly daylight and partly night. Furthermore, "noon," which originally meant the ninth hour of daylight, and then came to mean the sixth hour, is now number twelve. Talk about the time being out of joint.

Until the middle of the seventeenth century, there were no clocks capable of measuring small divisions of the hour. Nevertheless, the habit was established of dividing each hour into 60 minutes, and each minute into 60 seconds. This, too, began with the Sumerians, who applied this system to the division of each degree of arc into 60 minutes of arc and each minute of arc into 60 seconds of arc. The number 60 was chosen, like the number 12, because of the conveninece of its having many divisors. The number 60 is divided evenly by 2, 3, 4, 5, 6, 10, 12, 15, 20, and 30.

The solar day is defined as being 24 hours long; that is, 24 hours, 0 minutes, 0 seconds. The sidereal day, which I mentioned earlier in this chapter, is not quite so long. Actually, it is 23 hours, 56 minutes, 4 seconds long.

The difference is 3 minutes, 56 seconds.

Why ever should the sidereal day be shorter than the solar day by that queer amount? When the earth completes one turn, it completes one turn, whether you're making it by the stars or the sun, doesn't it?

The answer is, No! It makes a difference.

You see, the earth is not only rotating about its axis. It is also revolving about the sun.

As the earth moves in its journey about the sun, the stars are not measurably affected. They are so far away that the earth's orbit about the sun, which is 186 million miles across and seems to us to be enormous, is to the distant stars, to all intents and purposes, a point. Therefore, the earth might be viewed as rotating on its axis, but to be otherwise stationary with respect to the stars.

The sun is much closer to us than the stars are, however, and so it seems to shift position against the stars as earth moves about it.

At a given time, we see the sun against the stars in

a certain portion of the sky. (The stars in the immediate neighborhood of the sun cannot ordinarily be seen, of course, but we see the stars to its west just before sunrise, and those to its east just after sunset, and if we know the sky well, we will know the stars in between, which are in the immediate neighborhood of the sun.)

Half a year later, we are on the other side of the sun and, therefore, see it against the stars in the opposite side of the sky. Another half year and we are back where we were and the sun is back where it was. In other words, the sun seems to make a complete circuit of the sky in one year, or 365.2422 solar days.

That means that when the earth turns on its axis once with respect to the stars, the sun has moved a trifle eastward against the stars and the earth must continue to turn for 3 minutes and 56 seconds to catch up with it. Each day it has to make that small extra bit of turn to catch up with the sun, and after a whole year, earth has made a complete additional turn about its axis in order to keep up with a sun that has made a complete turn about the sky.

Therefore, while a year consists of 365.2422 solar days, it consists of 366.2422 sidereal days. The difference of 3 minutes and 56 seconds between a solar day and a sidereal day is $1/_{366.2422}$ of a year.

The sidereal day is the true period of earth's rotation relative to the universe in general, but there is no use arguing that point to anyone but astonomers. The people of earth are tied to the sun and to us what counts is when the sun (not Sirius, or the galactic center, or some distant quasar) crosses the meridian. For that reason, if you ask anyone how long it take the earth to turn on its axis, you will told it takes 24 hours. If you try to insist it takes 23 hours, 56 minutes, and 4 seconds, you'll very likely be hit with a brick.

Yet despite everything I've said, the interval from noon to noon is not exactly 24 hours. It is usually a little bit less or a little bit more than that. There are two reasons for this.

In the first place, the earth does not travel about the sun in a perfect circle. If it did, it would move always at the same speed, but it doesn't. The orbit is slightly elliptical, so that for half the year the earth is a bit closer that average. During the other half, it is a bit farther than average from the sun and moves about it at a speed a bit lower than average.

Standing on earth's surface, we see this terrestrial motion reflected in the apparent eastward motion of the sun against the stars. For half the year, this apparent motion is faster than average. That means that as earth's rotation carries the sun from east to west against the sky, its additional motion eastward because of its higher speed brings it to the meridian point a little later than it would have gotten there if earth had a circular orbit.

Then the sun begins to slow its apparent motion and its gain decreases and becomes a loss. Eventually, the loss decreases and becomes a gain. One can plot the time the sun crosses the meridian from day to day. There is a small bulge upward and a small bulge downward, but at the end of the year things are right where they are supposed to be. The difference is only a matter of a few minutes at the worst.

A second cause of irregularity rests with the fact that the earth's axis is tipped by 23.5 degrees to the plane of its revolution about the sun. At the equinoxes (March 20 and September 23), the sun's apparent motion across the sky cuts the equator at an angle and it moves more slowly from west to east. At the solstices (June 21 and December 21), it moves parallel to the equator and at a distance from it and seems to move more rapidly. In between the equinoxes and the solstices, the apparent motion slows or hastens. Again there is a bulge and a dip in the course of the year, but these even out by the end of the year.

If you add the two effects together, you have what is called "the equation of time."

Each of the two individual effects is symmetrical, with the highest part of the bulge and the lowest part of the dip equal in size and coming just six months apart. However, the two effects are not equal in size to each other and don't come at the same time of year. The equation of time, which is the sum of the two, is therefore asymmetric. It has two bulges upward in the course of the year and two bulges downward, and the bulges are of different size.

If we start at the beginning of the year, the sun is crossing the meridian a little late. This discrepancy increases and reaches a peak on February 12, when it is a little over 14 minutes late. The sun then begins catching up and is on time on April 14. It then moves ahead and is 8 minutes early on May 20. It is on time again on June 20 and falls behind so that by August 4 it is 6 minutes late. On August 29 it is on time again and then moves ahead until it is a little over 16 minutes early on November 3. It then slows up, is on time again on December 20, and continues falling back to begin the process all over again with the new year.

This unevenness in solar motion, involving never more than a quarter of an hour discrepancy, doesn't affect the ordinary person, but it would be one royal pain in the neck for clockmakers if they tried to devise a timepiece that kept exact time with the actual motion of the sun through the year.

Instead, timekeepers pretend that there is a sun crossing the meridian every day at the same time, as there would be if the earth's orbit were circular and its axis were not tipped. This is called the "mean sun"; "mean," from the Latin *median*, has the meaning of "average."

There is, therefore, "solar time," the time marked off by a sundial, and "mean solar time," in which the interval from noon to noon is always exactly 24 hours, whatever the sundial says.

You can mark out the position of the real sun east and west of the mean sun, east when it is fast and west when it is slow. At the same time, you can mark out the position of the real sun north and south of the equator (a position that varies in the course of the year due to the tilting of the axis).

The result is an asymmetric "figure eight," with the southern loop longer and wider than the northern loop (reflecting the asymmetry of the equation of time).

This asymmetric figure eight is called an "analemma," from a Latin word for "sundial," since it can be obtained in part from a comparison of the sundial noon with the clock noon. On large globes, it is placed in the middle of the Pacific Ocean, possibly because the region is empty and seems to need ornament. Certainly I see no use for it, though I admit I had to study it carefully in the process of writing this chapter.

Two more items before I close. We can't use mean solar time without still further modification.

Suppose each community adjusted its noon to when the mean sun crossed the meridian of some central point in that community. That would be "local mean time." Railroads, however, found it impossible to prepare timetables when each community had its own time. Therefore, the notion of "standard time" arose, in which fixed bands of earth's surface were assigned the same time regardless of their precise local mean time (see "The Times of Our Lives," in *Science, Numbers, and I,* Doubleday, 1968).

Finally, it turns out that as the days grow long, people sleep through several hours of sunshine in the morning, and then stay up after sunset and consume energy in order to light hours of darkness. If people got up earlier in the summer half of the year and went to bed earlier, some of that energy would be saved.

Can you imagine the American government ordering everyone to wake up an hour earlier and go to bed an hour later just to save desperately needed energy? Why, the American people in their proud independence and individuality would rise as one person and denounce those Washington bureaucrats who tried to tell them when to rise in the morning.

So the government sets up "daylight-savings time" and shoves the clock and hour ahead. When it now says 7 A.M., it's really 6 A.M. The clock is lying and everyone knows the clock is lying. However—

While Americans would scorn to be slaves to the government, they are pathetically eager to be slaves to the clock. An earnest, well-meaning government may tell them to get up at 6 A.M. instead of 7 A.M., and get a Bronx cheer in return, but when a lying *clock* tells them to do so, up they get like good little boys and girls.

I'll leave it to you to work out the moral of the story.

# 13.
# THE DISCOVERY OF THE VOID

The most pleasant science fiction convention I ever attended was the 13th World Science Fiction Convention, held in Cleveland in 1955. It was a small convention (only three hundred attendees) and very friendly, and I was the Guest of Honor, which helped.

I was even younger then than I am now, which also helped, and a number of my good friends were there, all of them (by some curious coincidence) much younger and handsomer than they are now, and some of them, alas, much more alive then than they are now.

One of the wonderful people I got to know well at this convention was Anthony Boucher, who was then editor of *F & FS*. He was the master of ceremonies at the convention, a sweet and gentle man, who is now dead and is forever enshrined in the hearts of those who knew him.

I was surprised, then, when I mentioned another attendee at the convention to have the kindhearted Tony snap, "I don't like him."

That was surprising, for the person under discussion seemed like a very nice fellow and I had found no trouble in liking him (but then I have no trouble liking almost anybody). I said, "Why don't you like him, Tony? He seems like a nice fellow."

And Tony shook his head and said, "He doesn't drink."

My eyes opened wide. I didn't know that drinking was a requirement for Tony's approval. I said, troubled, "But, Tony, I don't drink, either."

"That's different," said Tony. "He *acts* as though he doesn't drink. *You* act as drunk as the rest of us."

Drunker, actually. All those lushes at the convention sober up now and then and go about scowling at the world, but I never sober up. That's because I don't depend on alcohol, or any chemical substance, to get me lubricated. Life is one long high for me, and, in particular, writing one of these essays is enough to elevate me even in moments of difficulty. (I once wrote three *F & FS* essays in a row, without stopping, in order to retain my equilibrium when by beautiful, blond-haired, blue-eyed daughter broke her ankle.)

So let's get on with it.


In ordinary life, we tend to think of air as nothing at all. If we look into a container that holds nothing but air, we describe it as "empty." In a way, there is some justice to this if we compare air to some of the other objects around us.

The densest material we know, under the standard conditions about us on earth's surface, is the metal osmium. A cubic centimeter of osmium has a mass of

22.57 grams, so that its density is 22.57gm/cm³. (For those of you who have difficulty visualizing metric measurements, 1 cubic inch is equal to 16.39 cubic centimeters and 1 ounce is equal to 28,349 grams. Osmium therefore has a mass of 13.04 oz/in³. However, I'm going to stay metric.)

In comparison, the density of air is about 0.00128 gm/c³, which is about $1/17,600$ the density of osmium. Under those circumstances, it is tempting to dismiss air as negligible.

The fact that air has mass after all, and that it is therefore attracted by earth's gravitational field and can be measured as having weight, was not established until 1643. In that year, the Italian physicist Evangelista Torricelli (1608–1647) showed that if a tube, open at one end, is filled with mercury and up-ended into a trough of mercury, not all of it pours out. A 76-centimeter column of mercury remains in the tube, held there indefinitely by the weight of air pressing down on the mercury in the trough.

The density of mercury is 13.546 gm/cm³, and that is 10,583 times the density of air. This means that a column of mercury suspended in a closed tube must be counterbalanced by a column of air 10,583 times as high as that of the mercury column. Since air pressure supports 76 centimeters of mercury, the air column must be 8.04 kilometers (or almost exactly 5 miles) high.

This was a revolutionary piece of information. Until then, it had been casually assumed that air extended upward indefinitely—certainly as high as the moon and, possibly, as high as the stars.

Thus, in early science fiction stories, people were

pictured as reaching the moon by being hurled upward in a water spout, or through the help of large birds hitched to a coach. Such methods would only work if air were universal.

Now, for the first time, it was understood that the atmosphere was strictly a local phenomenon and that it hugged the earth's surface closely and that beyond that was nothing. People had to accept the fact that between the earth and the moon (or, more generally, between any two bodies in the universe) there was a larger or smaller gap of nothing at all. The only known way of crossing such a gap is by making use of action and reaction, as in the rocket, a principle first expounded by the English scientist Isaac Newton (1642–1727) in 1687.

In a way, then, Torricelli's experiement resulted in the discovery of space. To be sure, the entire universe, including the earth and you and me, is embedded in space. What is usually meant by the word, however, is the region beyond earth's atmosphere where there is, essentially, nothing, and which is distinguished from space generally by being referred to as "outer space."

An alternate word would be "void," which like "vacuum," refers to emptiness and which, for the purposes of this essay, I prefer. Torricelli's experiment, then, resulted in the discovery of the void.


How void, however, is the void? Is it empty? completely empty?

For instance, the atmosphere isn't actually just 5 miles high. That would be true if the density of the atmosphere were the same all the way up, but it can't be. That could be deduced from the fact that, in 1662, the British scientist Robert Boyle (1627–1691) showed that gases were compressed and made denser when placed under pressure.

The bottom of the atmosphere, in which we move, breathe, and have our beings, is compressed by the miles of air lying above it, so that we live in a sea of gas that is considerably denser than it would be if not for that pressure. As one moves upward in the atmosphere, there is a steadily smaller weight of it lying above, and, therefore, a steadily smaller air pressure pushing downward. For that reason, the air grows steadily less dense with increase in height. As it grows less dense, it spreads outward and upward and attains much greater heights than it would were the density constant everywhere.

Thus, at the top of Mount Everest, which is 8.8 kilometers high, the atmospheric density is only about three-eighths what it is at sea level. That is barely enough to allow our breathing apparatus to pump enough oxygen into our lungs for life to continue. As far as its practical use to ourselves and to other living creatures is concerned, then, we might consider the atmosphere to be only 9 or 10 kilometers high.

Nevertheless, the atmosphere extends farther upward, becoming ever less dense, ever less capable or supporting active life (though seeds and spores of various kinds might survive). To follow its upward extension, let's look at the atmosphere in a different fashion.

Of a given volume of clean, dry air, 78.084 percent is nitrogen, which consists of nitrogen molecules, each of which is made up of two nitrogen atoms ($N_2$). Next, 20.947 percent is oxygen, which consists of oxygen molecules, each of which is made up of two oxygen atoms ($O_2$). Then 0.934 percent is argon, which consists of individual argon atoms (Ar). Finally, 0.032 percent is carbon dioxide, which consists of molecules that are made up of one carbon atom and two oxygen atoms ($CO_2$).

These four components, taken together, make up 99.997 percent of the atmosphere. There are perhaps a dozen other trace components crowded into the remaining 0.003 percent of the volume, but we can ignore them.

We know the mass of individual argon atoms and the mass of individual molecules of oxygen, nitrogen, and carbon dioxide. Since we also know the mass of a cubic centimeter of air, we can calculate how many particles (a term we can use to include both atoms of argon and molecules of the other gases) are present in a cubic centimeter of air under standard conditions. The number is 26,880,000,000,000,000,000,000, or nearly 27 billion billion.

At the top of Mount Everest, the number is still about 10 billion billion per cubic centimeter, and we can make do with that—just barely.

At 100 kilometers above sea level, the atmosphere is less than a millionth as dense as it is at sea level, which makes it an extremely good vacuum by laboratory standards, but that means there are still 10,000 billion particles per cubic centimeter.

At 3,000 kilometers above sea level, the atmosphere is less than a millionth of a billionth of what it is at sea level, but that still means 10,000 particles per cubic centimeter. Even at 30,000 kilometers above sea level, almost one-twelfth of the way to the moon, there are still 10 particles per cubic centimeter.

You can see that the wisps of gas grow ever thinner but do not necessarily ever decline to actual zero over extended space. It might get down to 1 particle per cubic centimeter, or 1 particle per cubic meter, and yet never get to actual zero. The void is never *entirely* void, in other words.

But there is no use in seeking perfection. We might arbitrarily set some lower limit of density in defining an atmosphere, and where the density is still lower we might call it the void. Thus, about the highest phenomena for which earth's atmosphere might be considered responsible are the aurorae, some of which can be 1,000 kilometers high, at which height there are only about 300,000 particles per cubic centimeter. Let's call anything less than that "the void," not because it is absolutely empty, but because it is empty enough.

Under those conditions, all of space is void except for the utterly trifling volume in the immediate area of large bodies.

Every star has an atmosphere, of course, as our sun does, and every gas giant planet has an atmosphere, too, as Jupiter, Saturn, Uranus, and Neptune do. However, any object smaller than a gas giant rarely does. In our Solar System there are only four bodies that are smaller than the gas giants and that are known to have atmospheres. They are Venus, Earth, and Mars among the planets, and Titan among the satellites.

It was not long after Torricelli's experiment showed

the limited nature of earth's atmosphere, in fact, that astronomers began to realize that the moon, for instance, had no atmosphere.

Let's ask a question. Why does argon exist as single atoms, while oxygen and nitrogen pair up and form two-atom molecules. Without going into quantum-mechanical detail, let's just say that the arrangement of electrons around the argon atom is a very stable one. The stability cannot be decreased by having an argon atom share some of its electrons with another argon atom or with an atom of any other kind either. Argon atoms remain isolated, therefore.

The arrangement of atoms around an oxygen or nitrogen atom, however, is not particularly stable. A considerable increase in stability can be achieved if one oxygen atom shares electrons with another oxygen atom, or if one nitrogen atom shares electons with another nitrogen atom.

In combining, the atoms give up the excess energy required to maintain the unstable configuration in their single condition. To split such pairs apart, that excess energy must be supplied once more and be squeezed into the molecule. This is not an easy task and simply does not happen spontaneously under the conditions of the atmosphere around us, so that oxygen and nitrogen molecules remain in being there indefinitely.

In order for two individual atoms to share electrons, however, they have to be very close together, so close that we might as well view them as colliding.

As it happens, that is no problem under ordinary atmospheric conditions.

Suppose that all the nitrogen and oxygen molecules in the atmosphere existed in single splendor. What would happen?

With oxygen and nitrogen two-atom molecules separated into single atoms, there would be something like 53 billion billion particles per cubic centimeter, all of them atoms. If the atoms were moving, each one would have to travel only 3.5 millionths of a centimeter (on the average) before colliding with another. Since the atoms would be traveling at an average speed of about 6,500 centimeters per second nearly 100 miles an hour), there would be nearly 200 million collisions per second. That means that in just a tiny fraction of a second all the individual atoms would find partners. Oxygen atoms and nitrogen atoms would become oxygen molecules and nitrogen molecules and the heat liberated would raise the atmosphere to incandescence.

As one goes higher, however, the atmosphere grows less dense. The particles per cubic centimeter are fewer and, therefore, more widely spread. A particular particle must travel slightly farther, and therefore for a slightly longer period, before experiencing a collision.

About 85 kilometers above sea level, a particle must travel a full centimeter, on the average, before colliding with another particle. About 600 kilometers above sea level, a particle must travel 10 million centimeters (62 miles!) before colliding. What particles there are in the void would hardly ever collide.

Well away from the surface of the planet, the energetic radiation of the sun, ultraviolet light and X-rays, can supply the energy to break up the oxygen and nitrogen molecules into individual atoms. (Such radiation is absorbed long before it can reach the lower reaches of the atmosphere to wreak havoc there.) The individual atoms do not find it easy to collide in the low densities of the void, so that the higher one goes in the atmosphere, the more likely it will be that single atoms will be encountered.

At very great heights, oxygen and nitrogen tend to vanish almost altogether, and hydrogen and helium are encountered instead. These occur in the lowest portion of the atmosphere in the merest traces. Five particles out of every million are helium atoms (He), with the most stable arrangement of electrons of any atom. Five particles out of every 10 million are hydrogen molecules, each made up of a pair of hydrogen atoms ($H_2$).

Hydrogen and helium are the least dense of all the gases and would tend to float on the other gases, if temperature differences didn't tend to mix the atmosphere. Their particles are the smallest and lightest of the atoms, so that they move the fastest and are the least likely to be held by some particular gravitational field. For both reasons there is a greater tendency for them to find their way to the top of the atmosphere, and to "leak" into the void, than is true of the other gases.

It also happens that hydrogen and helium are the most common elements in the universe. Of all the atoms in existence, it is estimated that 90 percent are hydrogen and 9 percent are helium, while all the other elements together make up the remaining 1 percent.

This may seem unbelievable when the vast earth itself, as well as the moon, Mars, Mercury, Venus, and so on are made up almost entirely of everything *but* hydrogen and helium. However, the sun and the gas giants are made up chiefly, or even almost entirely, of hydrogen and helium, and since these five objects make up some 99.9999 percent of all the mass of the Solar System, the nature of the chemical composition of all its other bodies, including earth, doesn't amount to a sack of feathers.

Back in ancient Greek times, when the philosopher Democritus (ca. 470–380 B.C.) was developing the atomic theory, he held that matter consisted of nothing but atoms. There existed, he said, only atoms and, between them, the void.

Once Torricelli's crucial experiment was understood and it was found that air did *not* fill the universe, it was possible to modify Democritus's view on a vastly larger scale. In the universe, it appeared, there existed nothing but stars and the void.

Certainly, to the unaided eye that seems to be true. One can see the stars, and otherwise there is only black sky that seems to contain nothing at all. With the telescope, apparently empty stretches of sky are found to be full of stars to dim to be seen by the unaided eyes, but these stars are separated by apparent emptiness. No matter how great the magnification of the telescope and how many stars can be detected, there are always empty spaces between them.

We might decide, then, that the only items of interest in the universe are the stars (and any attendant planets they might have) and that the void is, so to speak, totally void of interest. What can you say about nothing, or about almost nothing?

And yet within a few years of the invention of the telescope, objects were discovered in the void that seemed not to be stars.

In 1612, the German astronomer Simon Marius (1573–1624) reported a fuzzy patch of light in the constellation of Andromeda. Such fuzzy patches were quite different in appearance from the sharp points of light that were the stars, and came to be called "nebulae" (from the Latin word for "clouds"). The one discovered by Marius was called the "Andromeda nebula" for three centuries.

Then, in 1619, the Swiss astonomer Johann Cysat (1586–1657) found that the middle star in the "sword" of Orion was actually a fuzzy patch of light rather than a sharp point. That was called the "Orion nebula."

Such fuzzy patches multiplied as telescopes improved, and they were frequently mistaken for comets by overenthusiastic astronomers. The French astronomer Charles Messier (1730–1817), beginning in 1771, compiled a list of over a hundred objects that might fool comet hunters if they were not warned.

As it turned out, many of the objects on Messier's list were, after all collections of stars. The Andromeda nebula is not a cloud of dust or fog, but is a vast conglomerate of hundreds of billion of stars, located so far away that the individual stars vanish into a luminous haze. Such conglomerates are now called "galaxies" and we speak of the "Andromeda galaxy." Thirty-eight of the objects listed by Messier have turned out to be galaxies.

Other objects on Messier's list are objects in our own Milky Way galaxy, but are "globular clusters" and "open clusters," collections of hundreds to hundreds of thousands of stars that blur together. There are fifty-eight such clusters on the list.

Then there are stars that have undergone some violent event and have emitted vast quantities of dust and gas that glow in the light of the star. These are "planetary nebulae" and a few of them are on the list. The very first item of Messier's list is the "Crab nebula" and that is what remains of a star that almost totally exploded as a supernova nine and a half centuries ago.

There are a few nebulae that really are glowing clouds of hydrogen and helium atoms, however. The Orion nebula is one of these. Two others are the "North America nebula" in Cygnus (so called from its shape) and the "Lagoon nebula" in Sagittarius (so called because it seems to consist of two parts with a dark channel or lagoon between).

The Orion nebula shines because in its vast volume are a number of hot stars that heat its gas and cause the hydrogen atoms to gain energy, lose their electrons, and become ionized. Such ionized hydrogen tends to give off its gained energy in the form of light. The atoms constantly regain energy from the stars within the nebula and as constantly radiate it away in a kind of fluorescent glow that is characteristic of these "emission nebulae."

It might seem astonishing that this glow can be seen across the vast distance that separates us from these nebulae. The gas of which they are composed is extremely rarefied, for they possess only from 1,000 to

10,000 particles per cubic centimeter. This is a density equivalent to that of our atmosphere at a height of 3,000 to 10,000 kilometers above sea level, and is low enough to make such nebulae meet our arbitrary definition of "the void." Still, when even such a thin scattering of atoms is spread over cubic light-years of space, it is enough to produce a visible glow.

There are thinner clouds, with only about 100 atoms per cubic centimeter, that are much more difficult to detect, since their densities are equivalent to our atmosphere at a height of 20,000 kilometers. Finally, the emptiest space, the voidest void, has only 0.3 particle per cubic centimeter (or about 5 particles per cubic inch).


Not all nebuale glow, of course.

When the German-British astonomer William Herschel (1738–1822) was studying the Milky Way, he noticed regions where there were very few stars, if any. These dark regions had definite boundaries, sometimes very sharp ones, and on the other side of the boundaries might be regions simply bursting with vast numbers of stars.

Herschel adopted the simplest explanation. He assumed that these dark regions in the Milky Way were really starless. That they were tunnels of emptiness boring through the crowds of stars and revealing the darkness of the void beyond the Milky Way. Earth seemed to be so situated as to allow us to look into the mouth of the tunnel. "Surely," said Herschel, "there is a hole in the heavens."

There are a number of such regions, however, and

with time, more and more were noted and described. By 1919, the American astronomer Edward Emerson Barnard (1857-1923) had catalogued the positions of 182 such dark regions, and by now the number known is well over 350.

It seemed to Barnard, and, independently, to the German astronomer Max F. J. C. Wolf (1863-1932), that it was very unlikely that there would be so many "holes" in the Milky Way with their openings all pointing toward the earth so that astonomers were able to peer into them.

It seemed much more likely that the dark regions were vast clouds of particles that contained no stars and did not, therefore, grow energetic and glow; they remained cold and dark. Such nebulae would block the starlight behind them, and would show up as dark blotches against the light that slipped past them on all sides.

These "dark nebulae" did not seem to be in any way the product of stars. Rather the reverse, for astronomers now believe that out of dark nebulae, stars might be formed under the proper conditions. The entire Solar System is thought to be the product of a dark nebula that, a little less than 5 billion years ago, condensed to form the sun and its planets.

If a dark nebula is large enough, many stars may form within it, and the first few of these would supply the energy to make an emission nebula out of it. In certain nebulae, such as the one in Orion, intensely dark and small circular patches are seen. These are called "Bok globules" after the Dutch-American astonomer Bart Jan Bok (1906-1983), who first studied them in the 1940s. They are thought to be clouds of gas that are actually condensing as we watch and that sometime soon (in an astronomical sense) will become newborn stars.

The dark nebulae are, like the emission nebulae, made up chiefly of hydrogen and helium and are about as dense, but from their very nature they show us they can't be made of gases only. If a dark nebula has 10,000 atoms of hydrogen and helium per cubic centimeter, it may very likely also contain 100 dust particles (each made up of tens, or hundreds, of atoms, including, perhaps, those of silicon and various metals) per cubic centimeter.

We know this must be so simply because a dark nebula absorbs sunlight. A dust particle is a hundred thousand times as effective in absorbing sunlight as an atom or molecule of gas is. We can see this in the case of our own atmosphere.

All the gaseous molecules in our atmosphere do very little in the way of absorption of sunlight, but allow some droplets of water or fragments of dust to enter and the condition changes at once. There may be very few bits of liquid or solid compared to the vast number of gas molecules present but those few bits produce a fog or mist that obscures the sunlight.

If only 1 percent of the particles in a nebula are dust and the other 99 percent are atoms and molecules of gas, then the dust still accounts for 99.9 percent of the obscuration of starlight.

Yet, although some nebulae emit light and some obscure light, and although both kinds of nebulae are very noticeable because of this, something much more subtle and fascinating takes place in them, and it is to this that I will turn in the next chapter.

# 14.
# CHEMISTRY OF THE VOID

I attended the Mystery Writers of America annual awards banquet earlier this year, along with my dear wife, Janet. We're a little sentimental about the function for we had first met, twenty-six years earlier, at one of those banquets.

In any case, I had been asked to hand out the Edgar for the best mystery novel of the year. Since this was the highest category of award, it came last and we sat patiently through the ceremony as a dozen successive speakers worked at being witty and clever.

Janet began to feel apprehensive. She knew that I was just a wee bit less than grateful for this opportunity to hand out a particularly important Edgar inasmuch as I myself had never been as much as *nominated* for anything by the MWA. She could also tell I was listening to all that wittiness and cleverness and was considering ways and means of topping them all.

So she leaned toward me and said, "Isaac, those poor nominees for the best novel have been in agonizing suspense all evening. Don't stretch it out. Just announce the five titles and authors and then read out the winner."

"Yes, dear," I said (I'm a remarkably well-behaved husband). "I will just announce the nominees and the winner."

And then came my moment and I bounded up to the podium in my usual youthful manner and read a sentence from the form letter I had that some of the names would be very difficult to pronounce and that if I had many problems, I was to call the MWA office in order that I might be coached on pronunciation.

I then folded the letter, put it in my pocket, said that I was proud of the multi-ethnic and pluralistic nature of American society, and would scorn to seek help. I would pronounce all the difficult names as best I could if the audience would bear with me.

I then turned to the list of five nominees, which happened, by the purest of coincidence, to include only authors whose names were particularly simple Anglo-Saxon in derivation. I read each book title, then hesitated over the name of the author, scanning it anxiously, then sounding it out with just a touch of difficulty and being rewarded each time with a roar of laughter. When I finished, and reached for the envelope that contained the name of the winner, I said sadly that it would probably contain the most complicated name of all, which I would then have to pronounce a second time. Sure enough, the winner was Ross Thomas, a name I pronounced with great difficulty. I got my sixth and loudest laugh, and returned to my seat.

"All I did was read the names, dear," I said to Janet.

Fortunately, as I write these chapters, there is no one at my elbow urging me to be brief, so I will now continue, in leisurely fashion, from the point where I left off in the previous chapter.

In the previous chapter, I talked about the void, the nearly empty spaces outside the immediate vicinity of large bodies. By earthly standards, the void is a vacuum and contains nothing—but not *altogether* nothing. It contains tenuous clouds of dust and gas, here and there. Even the clearest void, far removed from any star, must contain scattered atoms of one sort or another.

The question is: *What* sort or another?

Is there any way of analyzing an almost complete vacuum, at very long distance, in order to determine the nature of the thin, thin scattering of matter that it contains?

The beginning of an answer came in 1904. A German astronomer, Johannes Franz Hartmann (1865–1936), was studying the spectroscopic lines of a binary star, Delta Orionis. The two stars of the binary were too close to see as separate objects in the telescope, but as they swung about each other, first one would recede from us while the other approached, and then the other would recede as the first approached.

Both stars had spectroscopic lines, and when one receded while the other approached, one set of lines moved toward the red end of the spectrum, while the other moved toward the violet end. When the stars reversed their motion, so did the spectral lines. In other words, the spectral lines of the binary system would become double as the two stars swung about each other, then would merge as one star eclipsed the other, then would become double again in the other direction—over and over again.

But Hartmann noticed that one particular line *didn't move*. It was the line that represented atoms of the element calcium. The calcium can't be part of either star because both stars are moving. It has to be part of something that is stationary with respect to those stars and that has to be the faint wisps of interstellar gas that lie between the stars and the earth. Those wisps are extraordinarily tenuous but the number of atoms builds up in the light-years that separate the binary from ourselves and the starlight en route encounters enough of them to have the calcium wavelength detectably absorbed. Hartmann had, in effect, identified calcium among the components of interstellar gas.

Naturally, this wasn't accepted at once. There were other studies made, with conflicting results, and all sorts of competing theories were advanced. It was not till 1926 that the work of the English astronomer Arthur Stanley Eddington (1882–1944) showed convincingly that the interstellar gas explanation was correct. By that time, other types of atoms, such as those of sodium, potassium, and titanium, had also been detected in the interstellar gas.

These metals are relatively common elements on earth, and, presumably, in the universe in general. It was by then known, however, that hydrogen is by far the predominant element in the universe and should be predominant in the interstellar gas, too. About 90 percent of all the atoms in the universe are hydrogen and 9 percent are helium. Everything else put together makes up only about 1 percent at most. Why should one detect the minor constituents and not the overwhelming ones?

The answer is simple. Atoms such as calcium happen to absorb certain wavelengths of visible light quite strongly. Hydrogen and helium don't. Therefore, in studying the spectrum of visible light, the dark lines belonging to calcium and other such atoms in the void are detected. Nothing is seen in the case of hydrogen and helium.

Under one condition, hydrogen does become visible. A hydrogen atom consists of a nucleus with one positive charge that is cancelled by the negative charge of a single electron on the outskirts of the atom. The nucleus and electron together make up a "neutral hydrogen atom." If there is a hot star in the vicinity, however, the energetic radiation it releases tears the electron away from the nucleus, leaving a "hydrogen ion" behind. From time to time, the hydrogen ion recombines with the electron, giving off the spurt of energy that had been required to separate them, and this spurt can be detected.

Such hydrogen ion emissions were noted in luminous nebulae and could also be used to study the hot young stars in which the spiral arms of galaxies were rich, since the intense radiation of those stars formed sizable amounts of ionized hydrogen for light-years about. In 1951, the American astronomer William Wilson Morgan (1906–    ) was able to map the curves of ionized hydrogen that marked out the spiral arms of our own galaxy, in one of which our sun is located. Until then, our galaxy had been assumed to have a spiral structure, but this was the first piece of direct evidence.

Hydrogen ions, however, were found in only certain spots of the Galaxy. By far the major portion of the Galaxy consisted of small, dim stars. The space between these consisted of a thin gas of neutral hydrogen atoms that were invisible as far as ordinary light spectra were concerned. However, even as ionized hydrogen was being used to map the Galaxy's spiral arms, the situation with respect to neutral hydrogen atoms changed.

The German army had occupied the Netherlands in 1940 and had placed the land under the dark shadow of Nazi tyranny. Ordinary astronomical research became impossible, and a young Dutch astronomer, Hendrik Christoffell van de Hulst (1918–      ), was forced to see what he could do with nothing more than pen and paper.

The neutral hydrogen atom can exist in two forms. In one, the electron and the nucleus spin in the same direction; in the other, they spin in opposite directions. The two forms have a slightly different energy content. A vagrant photon of starlight might be absorbed by the lower energy form, which would then be converted into the higher energy form. That higher energy form would spontaneously slip back into the lower energy form, sooner or later, and give off the energy it had absorbed.

In 1944, van de Hulst showed that the energy given off would be in the form of a microwave photon with a wavelength of 21 centimeters (which would be about a 40-millionth as energetic as visible light). Any single hydrogen atom would emit that 21-centimeter wavelength only every million years on the average, but there are so many hydrogen atoms in outer space altogether that, at any given moment, large numbers are giving off these microwave photons and these could, in theory, be detected.

Prior to World War II, however, the instruments for the detection of such weak photons were lacking.

But just before World War II radar was developed and, during the war, a vast amount of research was put into it. Radar, as it happens, works with microwave beams and, by the end of the war, a great deal of technology had been worked out for microwave detection. Radio astronomy had become practical.

Using the new techniques, the American astronomer Edward Mills Purcell (1912–    ) detected the 21-centimeter radiation in 1951. It was now possible to study cold interstellar hydrogen and gain a vast amount of new information about the Galaxy as a result.

What's more, the new techniques of radio astronomy could be used to detect still other components of the interstellar gas.

For instance, the singly charged nucleus of the ordinary hydrogen atom consists of one proton and nothing else. There are a few hydrogen atoms, however, with a nucleus consisting of one proton and one neutron. Such a nucleus still has a single positive charge but is twice as massive as an ordinary hydrogen nucleus. This more massive hydrogen atom is usually called "deuterium."

Deuterium, like ordinary hydrogen, has two energy states and in slipping from the more energetic to the less it emits a microwave photon with a wavelength

244

of 91 centimeters. This radiation was detected by American astronomers at the University of Chicago in 1966, and we now know that about 5 percent of interstellar hydrogen is in the form of deuterium. In that same year, a Soviet astronomer detected the characteristic microwave radiation of the helium atom.

The dozen most common atoms present in the universe (and, therefore, in interstellar gas), in order of decreasing abundance, are hydrogen (H), helium (He), oxygen (O), neon (Ne), nitrogen (N), carbon (C), silicon (Si), magnesium (Mg), iron (Fe), sulfur (S), argon (Ar), and aluminum (Al).

As I said earlier, hydrogen and helium, together, make up 99 percent of the atoms in the universe. If these are set aside, the ten other types of atoms I have listed make up more than 99.5 percent of all the other atoms. In short, less than one in every twenty thousand atoms in the universe is of a variety other than the dozen I have listed. They can be ignored in what follows.

Let us now consider whether it is possible that the atoms of interstellar gas can exist there as anything other than single atoms. Can two or more atoms combine to form a molecule?

To combine, the atoms must first collide, and the individual atoms in the interstellar void are so far apart that collisions take place only very rarely. However, collisions do take place and since the universe has existed in more or less its present condition for 10 to 14 billion years, there would in time have been many, many collisions and many molecules would have been formed. To be sure, molecules, once formed, must withstand further collisions with radiation and energetic particles that would tend to break them apart again, but the balance between formation and breakup may be such that at any moment there are a certain number of molecules in existence.

And what kind of molecules would those be? To begin with, we can eliminate all atoms but the twelve types I have listed. Any other kind would be far too few to be involved in the formation of molecules in detectable concentration. Of the twelve listed, we can leave out three, since atoms of helium, neon, and argon simply do not combine with other atoms under any known conditions. As for silicon, magnesium, iron, and aluminum, they are not likely to form small molecules but tend to add on more and more atoms of themselves, with others such as those of oxygen, to form dust particles.

Such dust particles make up only about 1 percent of the mass of interstellar gas, but their presence is unmistakable. The individual atoms and small molecules of interstellar gas do not absorb significant amounts of sunlight, so that outer space is, generally, transparent.

Dust, however, is strongly absorbent. A mass of dust will absorb a hundred thousand times as much starlight as an equal mass of gas. In those volumes of space where interstellar dust is moderately abundant, the stars that lie behind them (relative to earth) are dimmed and reddened. If the dust is abundant enough, the stars are hidden altogether and we have the ''dark nebulae'' I mentioned in the previous chapter. (Individual atoms of the types usually making up dust particles are to be found in space, either not having yet bound themselves to the particles, or having been knocked loose from them. These account for spectral lines like those first detected by Hartmann.)

If we are thinking of true molecules, then, and not dust particles, we must confine ourselves to five types of atoms: hydrogen, oxygen, nitrogen, carbon, and sulfur, in that order of decreasing abundance.

Would combinations of these atoms exist in detectable quantity? The answer was Yes, for some combinations actually radiate in the visible light region when losing absorbed energy, and these could be detected by ordinary spectroscopic means even as early as 1941. Three such combinations are the carbon-nitrogen combination, "cyanide" (CN), the carbon-hydrogen combination, "methine" (CH), and methine with an electron missing, ($CH^+$).

These three combinations would not exist on earth. They could form, yes, but they would be very active and would quickly combine with other atoms or molecules in the environment to form more complicated and more stable molecules. In the interstellar gas, however, collisions are so few that these unstable combinations have no choice but to remain in existence at least to some extent.

There are no other likely molecular combinations that are apt to radiate in the visible light region, so it seemed for a while as though astronomers had reached their limit. In 1953, however, the Soviet astronomer Iosif Samuilovich Shklovskii (1916–1985) pointed out that the oxygen atom was more common than either carbon or nitrogen, so that the oxygen-hydrogen combination, "hydroxyl" (OH), was sure to be more common than either cyanide or methine. It, too, is unstable and wouldn't exist on earth but should exist in the tenuous gas wisps of interstellar space. However, it would not give off visible light, but microwave photons, instead.

Calculations showed that hydroxyl would give off four different characteristic wavelengths of microwaves, and these would serve as its "fingerprint." In October, 1963, the fingerprint of hydroxyl was detected and astonomers therefore had the key to further identifications.

For instance, with hydrogen the most common, by far, of the components of interstellar gas, we can expect that about 99.8 percent of random collisions would involve two hydrogen atoms. That means that a hydrogen-hydrogen combination, the "hydrogen molecule" (HH, $H_2$), should be the most common molecule in space. In 1970, the characteristic microwave radiation of the hydrogen molecule was detected in interstellar gas clouds.

At the present time, thirteen different two-atom combinations have been detected in space. They are HH, CO, CH, $CH^+$, CN, CS, CC, OH, NO, NS, SO SiO, and SiS. The last two involve the silicon atom and those may be incipient dust particles. Notice also that six of the thirteen involve the carbon atom.

In the middle 1960s, astronomers did not seriously expect to detect any atom combinations in space containing three or more atoms. They were sure that these could be formed through occasional lucky hits of a two-atom combination with an atom of hydrogen, or (less likely) with some other type of atom, or (least likely) with another two-atom combination. Yet it seemed that combinations of three or more atoms could scarcely be formed in detectable amounts in this way even in gas clouds where the atoms were more thickly distributed than in interstellar space generally, and where collisions were more likely to take place.

In 1968, however, came the big surprise that revolutionized ideas on the subject and established the new science of "astrochemistry." In November of that year, the telltale microwave fingerprints of the water molecule ($H_2O$) and the ammonia molecule ($NH_3$) were detected. The water molecule, as you see, consists of three atoms and the ammonia molecule of four.

These molecules are very stable and are common on planetary bodies. Earth has whole oceans of water, and the gas giants have atmospheres rich in ammonia. The problem is, though, how such complicated molecules could have been formed in detectable quantities in interstellar gas clouds in which the necessary collisions are not likely to take place often.

By now, no fewer than thirteen different three-atom combinations have been detected in interstellar space, of which eight contain a carbon atom. In addition, nine different four-atom combinations have been detected, of which eight contain a carbon atom (the ammonia molecule itself being the only one that does not).

The latest count I have seen lists twenty-four combinations of more than four atoms, and every one of them contains carbon atoms. The largest is a thirteen-atom molecule made up of a string of eleven carbon atoms with a hydrogen atom at one end and a nitrogen atom at the other end.

The more complicated these interstellar molecules are, the greater the puzzle of their formation. For one thing, the larger the molecule, the more rickety it is and the more likely it is that it would be broken up by stray photons of starlight. The feeling is, however, that the dust particles that exist in the interstellar gas clouds serve to shield the forming molecules and make it possible for them to continue to exist.

Various scheme of different collisions under different conditions have been advanced, and calculations based on these assumptions have been used to work out the kinds and relative number of molecules that are formed. None of the calculations are on the nose, but some end up in the ballpark. The general conclusion is that interstellar chemistry is strange because of the very unusual conditions (as compared with those with which we are familiar) but is not illegal. That is, the chemical and physical laws followed in the formation of those large interstellar molecules are the same as those we witness on earth.

It is interesting that of the fifty-nine different molecules identified in space, forty-six contain carbon atoms, including all but one of the combinations possessing more than three atoms. It would seem that in outer space, in a near-vacuum state, and with conditions extremely different from those on earth, it is nevertheless the carbon atom and no other on which complexity builds. This supports the conclusion I reached, for instance, in my essay "The One and Only" (see *The Tragedy of the Moon*, Doubleday, 1973).

There seems to be no doubt among astronomers

that the fifty-nine different atom combinations so far detected do not include all that there are. There may be hundreds or thousands of different combinations in the gas clouds, but detecting them is a problem. Clearly, the more complicated the molecule, the more interesting it is—but the fewer the numbers formed, the more difficult they will be to detect.

For instance, it isn't hard to imagine that, hidden among the cubic light-years of a gas cloud, there may be traces, here and there, of simple sugar molecules or of amino acids. These traces, if collected over the entire vast volume, may amount to tons and tons, but spread out as they are, they may be undetectable in the foreseeable future.

There is the importance of working out exactly how the molecules we have already detected have been formed. If we can work out an acceptable scheme we may be able to calculate just what additional, more complicated, molecules may be formed. That may present us with some pretty startling possibilities.

The British astronomer Fred Hoyle (1915–    ), for instance, already suspects that molecules may be built up in the interstellar clouds that are complex enough to possess some of the properties of life, though he remains a minority of just about one in this.

Still, it does seem very likely that the makeup of the interstellar gas clouds is relevant to the formation of life, even if they do not contain life themselves.

Our Solar system condensed out of an interstellar cloud of dust and gas, and while the solid chunks and clumps that formed the earth must have been heated in the process to the point where complicated carbon compounds, if any existed, were destroyed, the early earth may have been surrounded by a thin remnant of gas that contained various organic molecules. Much of this gas may have been swept away by the early solar wind, but some of it may have entered earth's early atmosphere and ocean.

In other words, are we wrong in attempting to work out the origin of life on earth from scratch—from very simple molecules? Suppose earth started with at least some of the more complicated molecules and was at least partway along the road to life at the beginning.

The smaller bits of material in the Solar system may preseve these original molecules. There are carbonaceous chondrites, a kind of meteorite, that contain small quantities of amino acids and fatlike molecules, for instance.

Comets may have them, too. Indeed, Fred Hoyle feels that comets may be hotbeds of primitive life, and that even molecules as complicated as those of viruses may exist there. He has even suggested that a close brush with a comet may result in a strain of virus being deposited in earth's atmosphere, a virus that may be pathogenic and one against which human beings would have little or no defense.

Could this be the origin of the sudden pandemics that afflict the earth—as, for instance, the Black Death of the fourteenth century? Or one might suggest that since earth is supposed to have passed through the tail of Halley's comet in 1910, it may have picked up some viruses that finally multiplied into the cause of the great influenza pandemic of 1918.

I don't believe any of this for a moment, and I don't know of any scientist who goes along with Hoyle in his more radical speculations, but I'm surprised it has not served as a theme for science fiction stories yet.

Or (since I can no longer read all the science fiction stories that come out) has it?

# 15.
# THE RULE OF NUMEROUS SMALL

I frequently get letters that ask me questions under the assumption that (1) I know everything and (2) I run a free information service.

Nevertheless, I answer when I can because I hate to disappoint people, especially if they're thoughtful enough to enclose a stamped, self-addressed envelope. Notice that I say "when I can," for there are times when I know nothing about a subject, and there are other times when I do know the answer but it would take pages and pages to explain it properly.

Yet every once in a while I am amply repaid for my trouble when a question makes me think. For instance, a woman wrote recently, asking me to explain the difference between a star and a planet. I grinned and was about to write: "A star is a large body at whose core there are nuclear reactions, so that it glows with heat and light. A planet circles a star and is too small to develop nuclear reactions, so that it is dark and shines only by reflected light from the star."

Then, a little to my surprise, I started thinking.

Can the matter of stars and planets be dismissed so easily? So I decided to write an essay on the subject.

If you consider a particular class of substances that come in different sizes, it often appears that the smaller the size, the more numerous the individual objects. Thus, stones are more numerous than boulders, pebbles than stones, sand grains than pebbles. Again, zebras are more numerous than elephants, mice than zebras, flies than mice, bacteria than flies.

"The Rule of Numerous Small" (as I call it) seems to apply to astronomic objects as well. The first indication of this came in connection with the brightness of stars. The ancient Greek astonomer Hipparchus divided stars into six classes—first magnitude for the brightest, then second magnitude and so on, down to sixth magnitude for the faintest stars. The number of first magnitude stars was small, but there was a larger number of second magnitude stars, a still larger number of third magnitude stars, and so on. Fully half of all the visible stars are of the sixth magnitude.

It seemed natural, in ancient and medieval times, to suppose that the visible stars were all there were. After all, if you don't see something, it's not there. With the invention of the telescope, however, it became apparent that there were stars too dim for the unaided eye to see. It was then possible to extend the line of magnitudes in the direction of dimness, to classify stars as of the seventh magnitude, the eighth, and so on. As it turned out, the number of stars at a particular magnitude continued to increase as one went down the line to dimmer and dimmer levels.

The ancients, of course, assumed that all the stars were at the same distance from us, since all, they thought, were attached to a solid celestial sphere. It seemed, therefore, that one star was dimmer than another only because it was the smaller of the two. (It was for this reason that the classes were called "magnitudes," a term that implies size rather than brightness.) For small stars to be more numerous than large stars did not seem at all strange.

Nowadays, however, we know that stars can be at widely different distances from us, and that a star can appear dim not only because it is small, but because it is distant.

It is possible, nevertheless, to determine the distances of various stars and to make allowance for that. We can then determine what the magnitudes would be if all were at the fixed distance of 10 parsecs (or 32.6 light-years). This gives us "absolute magnitudes." If we line up the stars in this fashion, we find that the larger the absolute magnitude and the lower the true brightness (or "luminosity") of a star, the smaller its mass and the more numerous the members of its class are. Thus, for every star that is more massive than the sun, and therefore more luminous, there are twenty stars that are less massive and less luminous than the sun.

Luminosity increases and decreases with mass, but much more rapidly. Thus, Procyon is 1.8 times as massive as the sun, but 5.8 times as luminous; Sirius is 2.5 times as massive as the sun, but 23 times as luminous. On the other hand, 70 Ophiuchi A is 0.95 times the mass of the sun, but only 0.36 times as luminous.

As mass continues to decrease, there must soon come a point where a star is too dim to be detected, and that means we are nearing a dividing line between stars and planets. What, then, is the least luminous (and, therefore, least massive) star known?

In my book *Alpha Centauri, the Nearest Star,* published in 1976, I list the least luminous star as "Van Biesbroeck's Star," so called because it had been discovered by a Belgian-American astonomer, George van Biesbroeck, about 1940. It can also be called by the more convenient name of VB 10.

The most recent value I can find for the absolute magnitude of VB 10 is 18.6. This means that VB 10 is 13.9 magnitudes dimmer than the sun. Magnitude is a logarithmic function, and for each unit of magnitude the luminosity must decrease by a factor of 2.512. It follows that VB 10 is only $\frac{1}{350,000}$ as luminous as the sun (i.e., 0.000003 S).

If our sun were replaced by VB 10, we would see, of course, a much smaller object in the sky, for VB 10 probably has a diameter of not more than 200,000 kilometers. This is about $\frac{1}{7}$ the diameter of the sun, so that VB 10 would appear to have an angular diameter of a little over 4 minutes. We would just make it out as a tiny disk rather than as a mere point of light.

It would be a deep red color for, considering its size, it would not develop enough nuclear energy at the center to raise its surface to more than red heat. VB 10's brightness would appear to be only 1.3 times that of the full moon now, so that earth would be bathed in nothing more than a ruddy moonlight. As for the moon itself, under such circumstances, it would shine by reflecting the red light of VB 10, with a total brightness, at the full, that would be equal to that of a star such as Arcturus. This brightness would be spread thin over the entire face of the moon. I doubt that the moon could be seen at all, in such a case, without some sort of magnification.

Since my book was published, however, VB 10 has been dethroned. In 1981, a dimmer star was identified and, in 1983, a still dimmer one. The latter, which now holds the record, is LHS 2924 and it has an absolute magnitude of 20. That would make it only $2/7$ as luminous as VB 10, or about $1/1,200,000$ as luminous as our sun (0.0000008 S). If it were put in the position of our sun, it would have a brightness only $2/5$ that of the full moon under present conditions.

How massive are these very dim stars? The answer to that is very hard to determine with any degree of certainty, but the best estimate seems to make them 0.06 times as massive as the sun (or $1/17$ the mass of the sun, if you prefer fractions).

Now let's approach the matter from the other end. What is the most massive body we know that is not massive enough to develop enough heat of any kind to glow of its own light?

The answer to that is simple. The largest nonglowing object we know is the planet Jupiter, which is visible only by the relected light of the sun.

Jupiter has a mass nearly $1/1,000$ that of the sun, or 0.001 S. This means that LHS 2924 has a mass of about 60 times that of Jupiter (60 J). Somewhere in between 1 J and 60 J, then, is the dividing line between a star and a planet. It may not be a sharp dividing line, because factors other than mass (say, the chemical constitution of the object) may affect the ability of an object to generate light of its own.

Still, as a rule of thumb we might say that 10 J is the boundary line. Any object that has a mass less than 10 times that of Jupiter might be considered a planet, while any object with a mass more than 10 times that of Jupiter might be considered a star.

By The Rule of Numerous Small, we might take it for granted that there must be a far greater number of planets in the universe than there are stars, since planets are small and stars are large.

Judging from our Solar system alone, this is certainly so. Our Solar system contains only one body that is large enough to be a star—the sun. It also has countless numbers of dark objects orbiting the sun, ranging in size from Jupiter down to microscopic dust particles.

The four largest bodies orbiting the sun—the "gas giants" Jupiter, Saturn, Uranus, and Neptune—make up a little over 99 percent of the total mass orbiting the sun. Everything else, including earth and the other small planets, all the satellites, asteroids, meteoroids, and comets, make up the other 1 percent. An objective observer, viewing the Solar system, would conclude that it consisted of the sun, four planets, and a scattering of inconsiderable debris.

The smallest of the sun's gas giants is Uranus, which has a mass about 1/22 that of Jupiter. We might say, then, rather arbitrarily, that all objects with masses above 10 J are stars; that objects with masses of from 10 J down to 0.05 J are planets; and that objects with masses of less than 0.05 J (including our earth) are "subplanets."

By this definition, we can say that our Solar system consists of one star, four planets, and innumerable subplanets. If other stars are attended by similar planetary systems (and the general feeling among astronomers is that it is likely that they are), then that alone should mean there are four times as many planets as there are stars.

This, however, perhaps unfairly restricts planets to dark bodies that orbit stars. Why should there not be planets that are totally independent of stars?

After all, stars are more numerous the smaller they are, in line with The Rule of Numerous Small, and why should we feel it right to limit ourselves only to those stars that can be sensed by our various instruments, any more than the ancients were right to limit themselves only to those stars that could be sensed by the unaided eye?

Whatever process is involved in star formation, that process seems to form middle-sized ones. Might that process not form very small ones—ones too small to develop nuclear reactions and to glow—more often still? Such very small "stars" would, in fact, be planets that did not orbit some nearby star but orbited the center of the Galaxy independently. These would be analogous to the asteroids of the Solar system, which are small enough to be satellites but aren't; instead of orbiting some nearby planet, they orbit the sun directly.

There is a tendency to call such independent planetary objects "black dwarfs," but I don't think that's a good name, since it is also used for white dwarf stars that have finally cooled off to the point where they no longer radiate detectably, and such black dwarfs can have masses far greater than those we associate with planetary bodies.

It seems to me that we ought to call planetary bodies that are independent members of the Galaxy "primary planets" and planetary bodies that orbit stars "secondary planets." (It may be that we ought to speak of primary and secondary subplanets, too.)

Although we have detected innumerable stars, we have not yet surely detected secondary planets outside the four in our own Solar system. There have been wobbles in the motion of some nearby stars that have been interpreted as suggesting the presence of secondary planets in orbit about them, but these suggestions are no longer generally accepted. More recently, belts of dust and gravel have been detected about some stars and these may suggest the presence of secondary planets, but we can't yet be certain of that.

As for primary planets, the situation seems much worse. After all, the only reason we can hope to detect secondary planets is precisely because there is a nearby star. The nearby star might wobble in its motion under the pull of a relatively large nearby planet, or the planet may be detected by its reflection of the light of that nearby star.

It's the essence of a primary planet (assuming that it exists) that there is not nearby star; no star to be made to wobble; no star to lend light for reflection.

Can we ever detect a primary planet, then, by direct observation?

Possibly!

Even if its gravitational field is too weak to be detectable and even if it neither gives off light of its own nor has available starlight to reflect, it may still be warm enough to give off infrared radiation or some characteristic kind of microwave radiation, and we may yet develop methods for detecting that.

The ability to do so may be enhanced in one of two fashions. We may establish a space telescope large enough to outdo the abilities of ground-based telescopes by a considerable margin; or we may develop starships that carry human beings on explorations far beyond the Solar system.

Finally, some primary planet may be moving about the galactic center in an orbit that intersects that of the sun. There may come a time when such a primary planet wanders in from interstellar space and moves through the outskirts of our planetary system. What excitement that would give rise to, except that the chances against such a thing happening would be— well, astronomical.

And yet there are other kinds of evidence altogether.

Judging from what we can see, the mass of a typical galaxy (like our own, for instance) might be 100 billion times that of our sun, and this mass is strongly concentrated toward the center. Perhaps 90 percent of it is in the central core of the galaxy, which makes up a small percentage of the total volume, while the remaining 10 percent is spread through the voluminous outer region.

This bears a certain similarity to the Solar system, where most of the mass is concentrated in the central sun, and only a small portion is spread through the vast outer regions of the system.

If this is truly the structure of typical galaxies, then the rotation of the parts thereof should show similarities to the rotation of the parts of our Solar system. The farther a planet is from the sun, for instance, the more slowly it traverses its orbit because the intensity of the gravitational pull of the sun falls off with distance. In accordance with this, astronomers were quite sure that the farther a galactic region would be from the galactic center, the slower the rotational movement of the stars in that region.

In recent years, astronomers have measured the rotational rates of galactic regions at increasing distances from the center and they have found, to their astonishment, that this doesn't hold true. The rotation rate does not decline with distance in the manner it was expected to.

The conclusion, then, is that the mass of the galaxies is *not* as heavily concentrated toward the center as had been thought. Instead, the mass must be spread out much more evenly—and well beyond what seems to be the edge of the galaxy.

It may be, then, that each galaxy (including our own), in addition to the stars we can plainly see, must have a halo of substantial mass enveloping the entire body of the galaxy, a halo made up of something that we *can't* see. And each galaxy must then be substantially more massive than we thought it to be.

If this is so, then another problem might be solved. Galaxies exist in clusters of various sizes. If a typical cluster of galaxies is examined, the individual galaxies are all found to be in random motion within the cluster. These motions would tend to break up and disrupt the cluster, unless the overall gravitational field of the cluster were intense enough to hold the components together despite their motion. However, the mass of a cluster, judging from its content of visible stars, is insufficient to hold it together even though it is obviously *being* held together. And the larger the cluster, the more the gravitational field of its visible stars falls short.

The matter becomes less puzzling at once if you take into account the mass of the invisible halos, and if you assume also that there must be some mass distributed between the individual galaxies of a cluster.

Finally, the universe as a whole has only about 1 percent of the mass required to keep it from expanding forever (that is, from being "open"); at least if one goes by the stars one can see in the universe. Some astronomers feel that it would make more sense to have the universe "closed"; that is, to have the expansion slowed to an eventual halt by the universe's overall gravitational field, and for there to follow a slow, but accelerating contraction, ending in a Big Crunch. Again the halo of the galaxies might supply the additional mass required for that.

But if the puzzles of the rotating galaxies, the held-together clusters, and the apparently open universe are all solved by the galactic halos, that simply supplies us with another puzzle. Of what does the halo consist? If there is mass there, which we can't see because it is not composed of stars, of what *is* it composed? (Astronomers call this the "mystery of the missing mass.")

One possibility, obviously, is that the mass of the halo consists of innumerable primary planets. Such objects would neither glow, nor find light to reflect, so they would be completely invisible to us. Their individual contributions to the gravitational fields of the galaxies, however, and to that of the universe as a whole, would be significant.

Suppose that the average mass of a primary planet was that of Jupiter. If there were a thousand such primary planets in the halo for every visible star in the Galaxy proper, then that would suffice to double the apparent mass of the Galaxy.

If you add in the primary planets scattered through the body of each galaxy and through the space between the galaxies, you might end with a hundred thousand primary planets for every visible star in the universe. That would explain the manner in which clusters of galaxies hang together, and would also suffice to close the universe, and to solve the mystery of the missing mass altogether.

To be sure, a hundred thousand primary planets for every visible star seems to be pitching it a bit strong even for The Rule of Numerous Small. But then, why blame all the missing mass on primary planets? There are other possibilities, too.

The galactic halos, and the space between the galaxies, might, for instance, be littered with black holes, which could each have a mass equal to that of a star, even to that of a giant star, and, indeed, even to that of a cluster of stars. Despite their potentially huge masses, black holes in isolation in space would be as invisible as primary planets.

It could be, then, that the halos are made up of substantial numbers of black holes with a correspondingly much smaller (and more believable) number of primary planets.

But, in that case, there is another puzzle. As galaxies formed, the pull of their own gravitational fields must have acted to concentrate their visible stars strongly toward the center. If so, why should it not have acted to concentrate primary planets and black holes strongly toward the center, also? Why should one type of mass have been concentrated and another kind not?

There is an even more serious objection. There are theoretical reasons for arguing that the number of protons and neutrons that can possibly exist in the universe as it is, is just about large enough to make up the mass that we can see. If, then, the mass of the universe is significantly greater than the mass we can see, that excess must be made up of something other than protons and neutrons.

Primary planets and, for that matter, black holes are made up almost entirely of protons and neutrons (as are the stars), so, if the theoretical argument is correct, primary planets and black holes cannot be responsible for missing mass. Astronomers are therefore looking for exotic explanations, such as neutrinos (see "Nothing and All" in *Counting the Eons*, Doubleday, 1983) or even more outlandish particles.

Even that, of course, does not mean that primary planets do not exist at all—merely that they do not exist in great numbers. There can still be a relative few without overrunning the permissible numbers of protons and neutrons. Of course, the fewer there are, the more difficult it would be to detect them.

* * *

But we have to ask ourselves another question. Does The Rule of Numerous Small always work?

Obviously not. If we consider human males, or human females, there are more, in each case, of medium size than of large size. But there are also more of medium size than of small size. In these cases, if one starts with very large individuals and considers the number of those who are smaller and still smaller, at first the numbers increase—but then they peak and begin to decrease again.

Is it possible that the size of stars also peaks at some value and that below that value the number of stars falls off precipitously?

Stars are formed through the condensation of huge clouds of gas and dust. In general, the more massive a could is, the more massive the star it will form, or the greater the number of stars it will form, or both.

Presumably, then, stars of very small mass are formed from relatively small clouds. But the smaller the cloud, the weaker its overall gravitational field and the less likely it is to undergo condensation under the inward pull of that field.

Some astronomers argue that a cloud that is so small that it can form only a primary planet on condensation, would be too small to condense at all. To be sure, secondary planets such as Jupiter, and secondary subplanets such as earth, have obviously formed, but they did so in the turbulent outskirts of a cloud that was large enough to form the sun on condensation.

From this point of view, it may be that primary planets are not at all likely after all. In that case, we may have to be satisfied with the simple distinction between stars and planets that I began with. Stars are massive and give off light. Planets are small, do not give off light and orbit stars.

That leaves us one final matter to take up before I am done.

In normal stars, like our sun, the energy that keeps it shining originates from nuclear fusion at its core, fusion that converts hydrogen-1 to helium-4.

For this to take place, however, a certain critical temperature must be reached at the core of the star as it condenses from the initial cloud. It has been calculated that if a condensing star is less than 0.085 times the mass of the sun (or about $1/12$ its mass), then that critical temperature won't be reached.

Yet a star that is somewhat less than $1/12$ the mass of our sun, once formed, may attain a central temperature that is high enough to fuse hydrogen-2 (deuterium) to helium-3. (Deuterium, of all stable atoms is the easiest to fuse.)

Deuterium, however, is much less common than hydrogen-1 and is quickly consumed past the point where it will serve as a fuel. Instead of shining for many billions of years as a small hydrogen-fusing star will, a deuterium-fusing star will shine only for a few million years.

An even smaller star may not reach a temperature that will bring about any fusion at all; yet the kinetic energy of its contraction may bring about a high enough temperature to make it glow—though for an even shorter period than is true of the deuterium fusers.

Such small stars, which produce light by means short of true hydrogen fusion, might not be considered by some to be true stars. Perhaps we might call them "substars."

But substars can be seen, if they exist, and are sufficiently close to us. Indeed, stars like VB 10 and LHS 2924 (and any other stars equally dim) seem to have masses somewhat less than $\frac{1}{12}$ that of our sun. In that case, they may well be substars.

# 16.
# SUPERSTAR

I'm a member of the Dutch Treat Club, all of the members of which are active, in some way, in the field of communications. (I write.) We meet once a week for lunch and conviviality. During the eight non-summer months, we also add a bit of entertainment, and some edification in the form of an improving lecture.

Once the entertainment failed and I got a hurry-up call the night before the meeting.

Could I stand up and entertain on this short notice?

Well, I can sing a little bit and I'm utterly unself-conscious, so I said, "Sure!"

Came the next day at lunch and when entertainment time arrived, I arose, and there was strong and instant suspicion among the audience. To make it worse, I cheerfully announced that I was going to sing all four stanzas of "The Star-Spangled Banner," even the third stanza, which had been officially eradicated for the crime of being too nasty to our good friends, the British, whom it describes, collectively, by making use of the loving expression "hireling and slave."

The Dutch Treaters do not stand on ceremony. They love our national anthem, but every single one of them was under the clear impression that he had heard it often enough in the course of ordinary life. There was no need to be "entertained" with it. I was therefore the recipient of loud groans and hisses.

I stood my ground unperturbed. I knew my Dutch Treaters. They could sing the first line of the first stanza, and knew an occasional additional phrase here and there. They were totally unaware of the existence of three more stanzas, however, and they knew nothing about the story behind the poem. I aimed to teach them.

I told them the stirring story. It dealt with the British three-pronged offensive of 1814 that threatened to destroy the young United States only thirty-one years after it had been recognized as independent by Great Britain. And the fate of America boiled down to whether Fort McHenry in Baltimore harbor would be taken or not, whether the night's bombardment by the British fleet would end with the star-spangled banner still flying over it—or not.

I prefaced each stanza with the necessary explanation and then sang it very clearly so that all the words could be heard. (I made life hard for the poor accompanist, however. I'm not a professional singer and I casually sang each stanza in a different key.)

When I finished the fourth stanza on a sustained and triumphant note, the same audience that had scoffed at the start rose in a spontaneous standing ovation such as I have rarely experienced. I am convinced that, in an excess of patriotism of an intensity they had never before experienced, those world-weary and jaded elderly gentlemen would have, one and all, marched to the nearest recruiting center and tried to enlist, if I had thought of suggesting it.

Afterward, when I thought the matter over, it seemed to me that my certainty that I could put the thing over arose out of my experience with these *F & SF* science essays. I am ready to discuss anything, however old hat it might seem to a reasonably sophisticated readership, simply because I am confident I can present it with an interesting slant.

Once, when I devoted a couple of essays to polar exploration, a reader wrote to tell me he suspected I had got it all out of some grade-school geography, but that he had found it good reading somehow.

Well, that's my job, so let's get on with it.

In the previous chapter, I talked about the smallest stars, so it is only reasonable that I should now deal with the largest stars.

I'll start with the sun, the only star close enough to be seen by the unaided eye (or by the telescope, for that matter) as anything but a dot of light.

By earthly standards, the sun is an enormous object. The earth's mean diameter is 12,742 kilometers, and if we set that value equal to 1, then the diameter of Jupiter, the giant planet of our Solar system, is 11.18. The diameter of the sun, however, by that same standard, is 109.2 (or 9.77 times the diameter of Jupiter).

If we consider the volume of the earth (just over a trillion cubic kilometers) as equal to 1, then the volume of Jupiter is just under 1,400. If Jupiter were hollow, 1,400 earths could be dropped into it, if they were all squashed tightly together. The volume of the sun, however, is just over 1,300,000 on that basis, so that if the sun were hollow, well over 900 Jupiters could be dropped into it.

One more thing. Let's set the mass of the earth (about 6 trillion trillion kilograms) equal to 1. In that case, Jupiter's mass is 317.83 and the sun's mass is 332,865.

The total mass of all the material moving about the sun—all the planets, satellites, asteroids, comets, and meteoric material—comes to 448.0 on the earth = 1 standard. That means that the sun's mass is 743 times all the rest of the Solar system put together. Another way of saying this is that the sun makes up 99.866 percent of all the mass of the Solar system.

But never mind comparing the sun with the planets. That's comparing a monstrous giant with insignificant pygmies. How does the sun compare with other stars? There, things might seem very different.

Let's start with the hundred nearest stars. They are close enough so that we are reasonably sure we know them all. If we tried to pick a hundred stars in some region that was relatively distant, the smaller ones might be too dim to see and we would end up with a skewed sample.

Of the hundred nearest stars, ninety-seven are distinctly smaller than the sun. One is about the same size as the sun and that is Alpha Centauri A, the larger of the Alpha Centauri double-star system.

Only two of the hundred nearest stars are more massive than the sun. One is Procyon, whose mass (if we set the sun's mass equal to 1) is 1.77, and the other is Sirius, whose mass is 2.31.

If the nearest stars are a fair sample of the whole (and they might be), then our sun is outdone in mass by only 2 percent of the other stars.

Does that mean that the sun is a monster star, and that we should look upon it as a giant?

No! We would be looking at things the wrong way.

Thus, there are only five bodies in the Solar system that are larger than the earth: the sun, Jupiter, Saturn, Uranus, and Neptune. Among the bodies smaller than the earth are four planets, several dozen satellites, a hundred thousand asteroids, a hundred billion comets, and countless trillions of meteoric fragments. Yet surely that doesn't mean that the earth is an object of monstrous size.

That there are so many objects smaller than the earth is only an example of "The Rule of the Numerous Small" that I discussed in the previous chapter. Despite the many smaller bodies, the existence of one sun is sufficient to make it clear that earth is a tiny object.

In the same way, it doesn't matter how few stars there are that are more massive than the sun. What counts is how much *more* massive than the sun some stars might be. If there are even a few that are much more massive than the sun, we would have to look upon the sun as a relatively small body.

Measuring the mass of a star is not easy. It is best done if the strength of its gravitational field can be measured, for that is proportionate to its mass. And that can be measured if there is a nearby body that responds to that gravitational field. From the nature of the response, we can determine the mass of the star.

Thus, in the case of binary stars, we have two stars circling a common center of gravity. If we know the distance of the binary, and can therefore calculate the distance between the two stars from their apparent separation, then we can use the distance of the binary and its period of revolution to obtain the total mass of the two stars. From the comparative size of the two orbits we can determine the mass of each.

Fortunately, more than half the stars in the sky are parts of binary systems. Procyon and Sirius are members of binary systems and they should be refered to as Procyon A and Sirius A, since each is more massive than its companion. In the case of these two stars, the companions, Procyon B and Sirius B, are white dwarfs.

For the moment, however, never mind mass. Another way of comparing stars is by how intensely then radiate. By this, I don't mean how bright they appear to be in the sky. That depends on their distance from us as well as upon the amount of radiation they deliver. A star can shine very brightly indeed and be so far from us as to be totally invisible even with the use of a telescope. On the other hand, a star can shine rather dimly but be so close to use as to make a brave show in the sky.

If we know the distance of various stars, however, we can make allowance for that distance, and calculate how much light each particular star would deliver if all were at the same standard distance from us. This level of brightness is called "luminosity."

The sun doesn't do badly as far as luminosity is

concerned. Of the hundred nearest stars, only two are distinctly more luminous than the sun and they happen to be the two that are also distinctly more massive: Procyon and Sirius. If we set the luminosity of the sun equal to 1, then the luminosity of Procyon is about 5.8, and that of Sirius is about 23.

Is this relationship between high mass and high luminosity significant? After all, there might be many reasons why a star could be particularly luminous. Luminosity might depend on chemical composition, on internal turbulence, on magnetic field intensity, on the rate of rotation, and so on. It might even be that several different properties of a star might each contribute and that luminosities would vary from star to star in a bewildering manner.

In 1916, Arthur Eddington began to work on this problem. He considered voluminous stars first. These were of low average densities and, he felt, considering their high surface temperatures, they might be gaseous throughout. There were certain "gas laws" established experimentally on earth, and the use of these gas laws might help explain what would happen to a volume of gas with the mass of a large star.

Eddington reasoned there would be one force pulling the gas together—gravitation. There would be two forces acting to keep the gas from being pulled together—gas pressure and radiation pressure.

As gravity pulls the star together, gas pressure rises, but so does gas temperature. In fact, following the gas laws, the temperature at the centers of stars must reach the millions of degrees. As the temperature goes up, the amount of radiation emitted, and, therefore, the radiation pressure, must go up, too, and very rapidly.

In the end, Eddington obtained an equation that related mass and luminosity. The higher the mass, the greater the gas and radiation pressure required to keep the star at an equilibrium size; and the greater the radiation pressure, the more brightly the star shone. It seemed that luminosity depended entirely on the mass of a star.

Eddington announced the mass-luminosity law in 1924 and, by that time, he found that ordinary stars like the sun, and even dwarf stars, also fit the relationship. From this, the conclusion was reached that all stars were gaseous throughout even when their average density was, as in the case of our own sun, equal to that of liquid water on earth, and when the density of the sun's center was much higher still.

The density of the center of the sun is about five times as great as that of platinum on earth. By Eddington's time, however, it was known that the mass of an atom is concentrated in the much smaller nucleus at its center. It was clear, then, that, under the pressures at the sun's center, atoms broke down and that the atomic nuclei moved about freely amid a sea of loose electrons.

The nuclei could be distributed much more thickly in such a sea than as part of intact atoms. For that reason, density could be very high and yet the freedom of motion of the ultratiny nuclei would be such that this "degenerate matter" would still behave as a gas.

Even white dwarfs, which are virtually all degenerate matter, behave as though they are gaseous. It is only when we get down to neutron stars that this rule fails and that we get a body as massive as a star that acts as though it were a solid.

Eddington's mass-luminosity law applies particularly to stars on the main sequence (stable, hydrogen-fusing stars like our sun). According to this law, luminosity varies at about the 3.5th power of the sun, give or take a bit. A star with three times the mass of the sun would have a luminosity of about fifty times that of the sun, and so on.

This has one important consequence that can be seen at once. The more luminous a star, and the more radiation it is emitting, the more hydrogen it must be fusing in order to produce that radiation.

Suppose a star is three times as massive as the sun. It has, then, three times the fuel supply. Since it is expending that fuel supply at fifty times the rate the sun does, it uses up its greater fuel supply in $3/50$ or, roughly $1/17$, the time the sun does.

A star need only fuse about a tenth of its hydrogen supply before its center begins to fuse helium. The star then leaves the main sequence and begins to expand into a "red giant." In a comparatively short time thereafter it will collapse into a white dwarf, neutron star, or black hole, depending upon its mass. A star with the mass of the sun will remain on the main sequence for about 10 billion years. (The sun's stay on the main sequence is now nearly half over, in other words.) A star with a mass three times that of the sun will stay on the main sequence only a little over half a billion years, because of the prodigal way in which it must fuse its hydrogen supply.

The more massive a star, then, the shorter its normal lifetime. The smallest stars on the main sequence will keep on dribbling out their radiation in small amounts for 200 billion years or more.

On the other hand, a star that is fifty times the mass of the sun will, by Eddington's mass-luminosity law, stay on the main sequence only 10,000 years, a mere eyeblink on the astronomical time-scale.

You can see, then, why there are so few stars more mssive than the sun. Not only are more massive bodies formed in fewer numbers than less massive ones by The Rule of Numerous Small, but those massive bodies that are formed vanish more quickly into collapse and dimness, and the more massive they are, the more quickly they vanish. If we can see a star right now that is fifty times the mass of the sun, we would expect, by Eddington's mass-luminosity law, that it was probably formed during historic times and that, in a few thousand years, it will have collapsed.

A second consequence of Eddington's law is that the greater the mass of the star, the greater the forces pulling inward and pushing outward, and the less leeway there is in the equilibrium. A little shift one way or the other in small stars would involve a relatively small excess of force. The star would quiver a bit, then come back to equilibrium. (The sun may have its quivers, but even though it is fairly massive, the quivers have never been enough to wipe out life on earth—and it wouldn't take much of a quiver to do that.)

On the other hand, as we consider stars that are more and more massive, little shifts involve larger and larger excesses of force. Eventually, just the normal quivers you might expect would be enough to drive a star into collapse or into explosion. Either way, it would no longer exist as a normal star. Eddington himself thought that a star that was some fifty times the mass of the sun was about as large as a star could get and still maintain a reasonable equilibrium. This might be called the "Eddington limit."

Here is a list of some notable stars in our own section of the Galaxy that are more luminous even than Sirius, and for each I have roughly calculated the mass on the basis of Eddington's law:

| Star | Luminosity (Sun = 1) | Mass (Sun = 1) |
|------|------|------|
| Pollux | 30 | 2.6 |
| Vega | 48 | 3.0 |
| Spica | 570 | 6.1 |
| Alpha Crucis | 910 | 7.0 |
| Beta Centauri | 1,300 | 9.5 |
| Canopus | 5,200 | 11.5 |
| Deneb | 6,300 | 12.2 |
| Rigel | 23,000 | 17.5 |

Far in the southern sky (invisible to people in the latitudes of Europe and of the northern United States) is the constellation of Dorado (the Goldfish). In that constellation is the Large Magellanic Cloud, the nearest galaxy to our own. We can see considerable detail in it, including a star more luminous than any in our neighborhood of our own galaxy. It is invisible to the unaided eye, but the Large Magellanic Cloud is 55,000 parsecs away. Allowing for that enormous distance, we can see that S Doradus must be 480,000 times as luminous as the sun and must have a mass

280

of about forty times that of our sun. It nears Eddington's limit.

It seems, then, that there are stars that may be fifty times the mass of the sun, and that the sun is, in turn, about ten times more massive than the dimmest stars. From that it would seem that our sun is, at best, only a middle-sized star and that is what it is usually considered to be.

There is a catch, though. Eddington's upper limit is undoubtedly too conservative. In 1922, two years *before* Eddington announced his mass-luminosity law, a Canadian astronomer, John Stanley Plaskett (1865–1941), had discovered that a certain apparently unremarkable star was a binary. It turned out that each star is from 65 to 75 times as massive as the sun, and each one may be about 2,500,000 times as luminous as the sun.

This binary, called "Plaskett's twins" (a more dramatic name than the official "HD 47129"), if put in place of the sun, would probably vaporize the earth in short order. Earth would have to circle Plaskett's twins at a distance 55 times the average distance of Pluto from the sun (that is, $1/100$ of a parsec) in order to reduce the total radiation received to that which we now get from the sun. And even so, it would kill us, for the light from Plaskett's twins would be much, much higher in ultraviolet and X-rays than the light of our sun is.

The existence of Plaskett's twins enforced a rise in Eddington's limit to a mass of about 70 times that of the sun, a limit given in *The Cambridge Encyclopaedia of*

*Astronomy,* an excellent book that was published in 1977.

During the 1970s, however, the physics of large stars was reworked, making use of knowledge gained since Eddington's time. Turbulence within a star, for instance, plays a much greater role than had been thought. Then, too, there is a continuous and appreciable loss of mass from large stars through stellar winds, a phenomenon unknown in Eddington's time.

Neither turbulence nor mass loss invalidates the mass-luminosity law (which, after all, is backed not only by theory, but by careful observation). They do, however, raise Eddington's limit to surprisingly high values, making it clear that the stability and life span of "very massive stars" are greater than had earlier been expected.

There had been reports of such very massive stars (or "superstars," as I like to call them) with masses more than 100 times that of the sun, but in view of Eddington's low limit, such reports were received with the greatest skepticism. Once, however, the theory had been adjusted to allow the existence of superstars, a number have been reported and it may be that one star out of 2 billion is a superstar of more than 100 times the mass of our sun. That means that there could be 100 or 150 superstars in our galaxy alone.

Some particular superstars have been identified. In my essay "Ready and Waiting" (in *The Road to Infinity,* Doubleday, 1979), I referred to a peculiar star, Eta Carinae, as unusually unstable and therefore, very possibly, the next supernova. At that time, I had not yet caught up to the notion of superstars (trying to keep up with all of science is particularly exhausting and incredibly frustrating), but now my impression is that Eta Carinae owes its peculiarities more to its being a superstar than to being a pre-supernova.

As long ago as 1970, there were reports that Eta Carinae might be a superstar. Now, a number of astronomers seem to agree on that and it may be that Eta Carinae has a mass no less than 200 times that of the sun. Its luminosity may be 5 million times that of the sun; that is, $10\frac{1}{2}$ times that of S Doradus, or equal to both Plaskett twins put together.

Eta Carinae is losing mass. In my earlier essay I took this to be a sign that it was a pre-supernova, but superstars always lose mass in the form of a good, brisk stellar wind. That helps keep them realtively stable. The fact that Eta Carinae's stellar wind contains nitrogen and oxygen, which again I took as a pre-nova sign, may only signify that superstars undergo strong inner turbulence, which may again serve to keep them stable.

The stellar wind may mean an annual loss to Eta Carinae of something like one full solar mass in a hundred years. If that were to continue unchanged, Eta Carinae would be entirely gone in 20,000 years, but, of course, it won't. As Eta Carinae loses mass and drifts out of its superstar status, its stellar wind is bound to decrease in volume. It may be that superstars, through their active stellar wind, slowly lose their hydrogen-rich envelope and become naked star centers that are primarily helium. These are called "Wolf-Rayet stars," after the astronomers who first remarked on them.

Another superstar in our own galaxy is thought to be one called "P Cygni." It is much like Eta Carinae, but smaller. Its mass is about half that of Eta Carinae and is perhaps nearly 100 times the mass of our sun. It is only about a third as luminous as Eta Carinae, but that still makes it about 1,500,000 times as luminous as the sun and over 3 times as luminous as S Doradus.

But what is the most luminous superstar known? Well, back to the Great Magellanic Cloud.

Within the cloud is a gaseous nebula something like the Great Orion nebula in our own galaxy. The nebula in the cloud is vastly greater, however. It covers an area of about 3,000 parsecs by 1,000 parsecs and is the brightest object in the Great Magellanic Cloud. It can even be seen with the unaided eye. It is larger than any nebula in our own galaxy, or, in fact, than any we can make out in any galaxy close enough to us to have visible detail. It is called the "Tarantula nebula" because its shape reminded some observers of a spider.

The Tarantula nebula seems to contain a number of Wolf-Rayet stars, which may be the descendants of a whole group of superstars. The nebula may, at least or in part, be the product of the blown-off outer portions of those superstars.

Some people think that almost all the luminosity and ionization of the Tarantula nebula now comes from a central area no more than $\frac{1}{10}$ of a parsec across. The area may contain several stars but, in 1981, one group of astronomers became convinced that it was the site of a single superstar, the most luminous we may have detected up to this point. This superstar is called "R136a."

R136a may have a mass that is possibly 2,000 times that of the sun. In mass, the sun is to R136a as the planet Mercury would be to a planet somewhat larger than Saturn. That makes our sun look like a pipsqueak indeed, but don't let that offend your solar chauvinism. Superstars make space unlivable for light-years about themselves, and what's so great about that?

R136a may be about 60 million times as luminous as the sun, which would make it 40 times as luminous as Eta Carinae. Its surface temperature may be as high as 60,000° K.

If earth were circling R136a, it would have to do so at a distance of $\frac{1}{26}$ of a parsec ($\frac{1}{8}$ of a light-year) to reduce the apparent radiation level to that of our own sun and even then we would have to live underground to avoid the hard radiation.

What it amounts to, then, is that we are now aware of a class of remarkable stars whose existence was not dreamed of as little as fifteen years ago and whose existence was, in fact, considered to be impossible. If we can now study such stars in detail, we may learn a great deal about stellar astrophysics that can then be applied to more ordinary stars, including our own delightful Pipsqueak.

[Note: Only a few weeks after this essay first appeared, further astronomical investigation seriously dimmed the possibility of superstars, particularly in the Tarantula nebula. Too bad!]

# 17.

# FAR AS HUMAN EYE COULD SEE

I received a communication from a tax-gathering department the other day, and such communications are marked by two unfailing characteristics. First, they are tremble-inducing (What are they after? What have I done wrong?). Second, they are written in High Martian. It is simply impossible to interpret what they are saying.

As nearly as I could make out, something was wrong with one of my minor taxes of 1979. I had underpaid by $300 and was being soaked that, plus $122 in interest and so the total came to $422. Somewhere among the rank and sprouting verbiage there was a collection of words that sounded like the threat of my being strung up by one big toe for twenty years if I didn't pay in five minutes.

I called my accountant, who, as always, was utterly calm at this threat to someone else's existence. "Send it to me," he said, stifling a yawn. "I'll look at it."

"I think," I said, nervously, "I had better pay it first."

"If you wish," he said, "since you can afford it."

So I did. I wrote out the check, put it in an envelope, and sprinted to the post office to make the deadline and save my big toe.

Then I took the document to my accountant, who used his special accountant's magnifying glass to study the small print. Finally, he was ready with his diagnosis.

He said, "They're telling you *they* owe *you* money."

"Then why are they charging me interest?"

"That's the interest *they* owe *you*."

"But they threaten me if I don't pay."

"I know, but tax collecting is a dull job and you can't blame them for trying to inject a little harmless fun into it."

"But I paid them already."

"It doesn't matter. I will simply write to them and explain that they terrorized an honest citizen and they will eventually send you a check for $844, covering their debt to you, plus your unnecessary payment." Then he added, with a jovial smile, "But don't hold your breath."

That gave me my opportunity for the last word. "A person who deals with publishers," I said, austerely, "is accustomed to not holding his breath for payment."*

And now, having established my credentials as a keen-eyed, farseeing individual, let us do a little keen-eyed farseeing.

* * *

---

*Actually, the tax people sent me back my check within ten days, saying they had no right to it.

Suppose I dip into the future far as human eye could see (to coin a phrase Alfy Tennyson once used). If I do that, what will I see happening to earth? Let us assume, to begin with, that earth is alone in the universe, albeit with its present age and structure.

Naturally, if it is alone in the universe, there is no sun to light and warm it, so its surface is dark and at a temperature near absolute zero. It is, in consequence, lifeless.

Its interior, however, is hot because of the kinetic energy of the smaller bodies that coalesced to form it 4.6 eons ago ("eon" being taken to be 1,000,000,000—one billion—years). The inner heat would escape only slowly through the insulating rock of its crust, and, besides, would be continually renewed by the breakdown of such radioactive constituents of earthly matter as uranium-238, uranium-235, thorium-232, potassium-40, and so on. (Of these, uranium-238 contributes about 90 percent of the heat).

We might assume, then, that earth, alone in the universe, would endure for a long time in its condition of cold outside and hot inside. The uranium-238, however, slowly decays, with a half-life of 4.5 eons. As a result half of the original supply has already disappeared, and half of what remains will disappear in the next 4.5 eons, and so on. In about 30 eons from now, the uranium-238 remaining in the earth will be only about 1 percent of the present content.

We may expect, then, that with time, the earth's internal heat will leak away and will be less and less efficiently replaced by the shrinking supply of radioactive materials. By the time the earth is 30 eons older

288

than it is now, it will be merely lukewarm inside. It would continue to lose heat (at a slower and slower rate) for an indefinite period, getting even closer to absolute zero, and, of course, never quite reaching it.

But the earth is not the only body in existence. In our Solar system alone, there are countless other objects of planetary and subplanetary size, from mighty Jupiter down to tiny dust particles and even down to individual atoms and subatomic particles. There might be similar collections of such nonluminous objects circling other stars to say nothing of such objects wandering through the interstellar spaces of our Galaxy. Suppose, then, that the entire Galaxy was made up of such nonluminous objects only. What would be their ultimate fate?

The larger the body, the higher the internal temperature and the greater the internal heat gathered together in the process of formation; and, in consequence, the longer the time it would take to cool off. My rough guess is that Jupiter, with a little over three hundred times the mass of the earth, would take at least a thousand times as long to cool as earth would—say, 30,000 eons.

In the course of this vast length of time, however (two thousand times the present age of the universe), other things would happen that would outweigh the mere process of cooling off. There would be collisions between bodies. In the periods of time we're used to, such collisions would not be common, but over the space of 30,000 eons, there would be very many. Some collisions would result in breakups and disintegrations into still smaller bodies. Where a small body collides with a much larger one, however, the smaller body is trapped by the larger, and remains with it. Thus, earth sweeps up trillions of meteorites and micrometeorites each day, and its mass slowly, but steadily, increases as a result.

We may consider it a general rule, in fact, that as a result of collisions, large bodies grow at the expense of small bodies, so that, with time, small bodies tend to grow rare, while large bodies grow ever larger.

Each collision that adds to the mass of a larger body also adds kinetic energy that is converted into heat, so that the cooling-off rate of the larger body is further slowed down. In fact, particularly large bodies, which are especially effective in gathering up smaller ones, gain energy at a rate that will cause them to warm up rather than cool down. This higher temperature, plus the greater central pressures that come with increasing mass, will eventually (when the body is at least ten times the mass of Jupiter) bring about nuclear reactions at the center. The body will undergo "nuclear ignition," in other words, and its overall temperature will rise even higher until, finally, the surface grows faintly luminous. The planet will have become a feeble star.

One can then imagine our galaxy as consisting of planetary and subplanetary nonluminous bodies that gradually develop, here and there, into faint specks of light. It is useless to do so, however, because in actual fact, the Galaxy, in forming, condensed into bodies massive enough to undergo nuclear ignition to begin with. It consists of as many as 300 billion stars, may of them quite brilliant and a few of them thousands of times as luminous as our sun.

What we must ask, then, is what will become of the stars, for their fate will far outweigh anything that will happen to the smaller nonluminous bodies that, for the most part, circle in orbit about the various stars.

Nonluminous bodies can exist without serious change (except for the cooling process, and occasional collision) for indefinite periods, because their atom structure resists the inward pull of gravity. The stars, however, are in a different situation.

Stars, being far more massive than planets, have far more intense gravitational fields and their atomic structure smashes under the inward pull of those fields. As a result, stars, on formation, would at once shrink to planetary size and gain enormous densities, if gravity were all that need be taken into account. However, the vast temperatures and pressures at the center of such massive objects result in nuclear ignition, and the heat, developed by the nuclear reactions at the core, succeeds in keeping the stars' volumes expanded even against the pull of their enormous gravities.

The heat of stars, however, is developed at the expense of nuclear fusion that converts hydrogen to helium, and, eventually, to more complicated nuclei still. Since there is a finite amount of hydrogen in any star, the nuclear reactions can continue only for so long. Sooner or later, as the content of nuclear fuel diminishes, there is a gradual failure of the ability of nuclear-developed heat to keep stars expanded against the inexorable and forever continuing inward pull of gravity.

Stars no more massive than our sun eventually consume enough of their fuel to be forced to undergo a rather quiet gravitational collapse. They contract to "white dwarfs" the size of earth or less (though retaining virtually all their original mass). White dwarfs consist of shattered atoms, but the free electrons resist compression through their mutual repulsion, so that a white dwarf, left to itself, will remain unchanged in structure for indefinite periods.

Stars that are more massive than the sun undergo more drastic changes. The more massive they are, the more violent the events. Beyond a certain mass, they will explode into "supernovas" that are capable of radiating, for a brief period, as much energy as 100 billion ordinary stars. Part of the mass of the exploding star is blown off into space and what is left can collapse into a "neutron star." To form a neutron star, the force of collapse must break through the electron sea that would maintain it in the form of a white dwarf. The electrons are driven into combination with atomic nuclei, producing neutrons which, lacking electrical charge, do not repel each other but are forced together in contact.

Neutrons are so tiny, even compared to atoms, that the entire mass of the sun could be squeezed into a sphere no more than 14 kilometers in diameter. The neutrons themselves resist breakdown, so that a neutron star, left to itself, will remain unchanged in structure for indefinite periods.

If the star is particularly massive, the collapse will

be so catastrophic that even neutrons will not be able to resist the in-pulling effect of gravity, and the star will collapse past the neutron-star stage. Past that there is nothing to prevent the star from collapsing indefinitely toward zero volume and infinite density and a "black hole" is formed.

The amount of time it takes a star to use up its fuel to the point of collapse varies with the mass of the star. The larger the mass, the more quickly its fuel is used up. The largest stars will maintain an extended volume for only a million years, or even less, before collapsing. Stars the size of the sun will remain extended for perhaps 10 to 12 billion years before collapsing. The least massive red dwarfs may shine up to 200 billion years before the inevitable.

Most stars of our galaxy were formed not long after the Big Bang some 15 billion years ago, but a scattering of new stars (including our own sun) have been forming steadily ever since. Some are forming now, and others will continue to form for billions of years to come. The new stars that will form out of dust clouds are, however, limited in number. The dust clouds of our galaxy amount to only 10 percent of its total mass, so that 90 percent of all the stars that can form have already formed.

Eventually, the new stars will also collapse and while the occasional supernova will add to the interstellar dust, the time will come when no new stars can form. All the mass of our galaxy will have been collected into stars that will exist in collapsed form only, and in three different varieties: white dwarfs, neutron stars, and black holes. In addition, there will be various nonluminous planetary and subplanetary bodies here and there.

Black holes, left to themselves, give out no light and are as nonluminous as planets. White dwarfs and neutron stars do give off radiation, including visible light, possibly even more per unit surface than ordinary stars do. However, white dwarfs and neutron stars have such small surfaces compared to ordinary stars that the total light they emit is insignificant. A galaxy composed of only collapsed stars and planetary bodies would therefore be essentially dark. After about 100 eons (six or seven times the present age of the Galaxy) there would be only insignificant sparks of radiation to relieve the cold and darkness that would have fallen everywhere.

What's more, what specks of light do exist will slowly diminish and vanish. White dwarfs will slowly dim and become black dwarfs. Neutron stars will slow their rotations and emit weaker and weaker pulses of radiation.

These bodies, however, will not be left to themselves. They will all still make up a galaxy. The 200 or 300 billion collapsed stars will still have the shape of a spiral galaxy and will still revolve majestically about the center.

Over the eons, there will be collisions. Collapsed stars will collide with bits of dust, gravel, even sizable planetary bodies. At long intervals, collapsed stars will even collide with each other (releasing quantities of radiation that will be large in human terms but insignificant against the dark mass of the Galaxy). In general, the tendency, in such collisions, will be for the more massive body to gain at the expense of the less massive body.

A white dwarf that gains in mass will eventually become too massive to remain one and will reach the point where it collapses suddenly into a neutron star. Similarly, a neutron star will reach the point of collapsing into a black hole. Black holes, which can collapse no further, will slowly gain in mass.

In a billion eons ($10^{18}$ years), it may be that our Galaxy will consist almost entirely of black holes of varying sizes—with a smattering of non-black-hole objects, from neutron stars down to dust, making up but a very small fraction of the total mass.

The largest black hole would be the one that was originally at the center of the Galaxy, where the mass concentration has always been the greatest. Indeed, astronomers suspect there is already a massive black hole at the Galaxy's center, one with a mass of perhaps a million suns, and steadily growing.

The black holes making up the Galaxy in this far future will be revolving about the central black hole in orbits of varying radii and eccentricity and two will every now and then pass each other comparatively closely. Such near misses might well allow a transfer of angular momentum, so that one black hole will gain energy and will loop farther out from the galactic center, while the other will lose energy and will drop in closer to the center.

Little by little, the central black hole will swallow up one smaller black hole after another, as the small ones lose enough energy to approach the center too closely.

Eventually, after a billion billion eons ($10^{27}$ years),

the Galaxy may consist essentially of a "galactic black hole," surrounded by a scattering of smaller black holes that are far enough away to be virtually independent of the gravitational influence of the center.

How large would the galactic black hole be? I have seen an estimate for its mass as a billion suns, or 1 percent of the total mass of the Galaxy. The remaining 99 percent would be made up (almost entirely) of the smaller black holes.

Yet I feel uneasy about that. I can't offer any evidence, but my instinct tells me that the galactic black hole should be more like 100 billion suns in mass, or half the mass of the Galaxy, while isolated black holes make up the other half.


Our galaxy does not, however, exist in isolation. It is part of a cluster of some two dozen galaxies, called the "Local Group." Most of the members of the Local Group are considerably smaller than our galaxy, but at least one, the Andromeda galaxy, is larger than ours.

In the $10^{27}$ years that would suffice to convert our galaxy into a galactic black hole surrounded by smaller ones, the other galaxies of the Local Group would each be converted into the same. Naturally, the various galactic black holes would vary in size according to the original mass of the galaxy in which they formed. The Local Group, then, would consist of two dozen or so galactic black holes, with the Andromeda black hole the largest and our Milky Way black hole next.

All these galactic black holes would be revolving

about the center of gravity of the Local Group, and various galactic black holes would undergo near misses with a transfer of angular momentum. Again, some would be forced far from the center of gravity and some would sink closer to it. Eventually, a supergalactic black hole would form that might have a mass (my guess) equal to 500 billion suns—a mass equal to about twice the mass of our own galaxy—with smaller galactic black holes and subgalactic black holes circling in enormous orbits about the supergalactic black hole, or actually drifting off in space, altogether independent of the Local Group. This should be a better picture of the situation after $10^{27}$ years than the one drawn earlier from our galaxy alone.

The Local Group is not all there is in the universe, either. There are other clusters, perhaps as many as a billion of them, some of them large enough to include a thousand individual galaxies or more.

The universe, however, is expanding. That is, the clusters of galaxies are receding from each other at large velocities. By the time $10^{27}$ years have passed and the universe consists of supergalactic black holes, those individual supergalactic black holes will be receding from each other at speeds that it is not likely they will ever interact significantly.

What's more, the smaller black holes that will have escaped from the clusters and will be wandering about through intercluster space are not likely to encounter major black holes in the ever expanding space through which they move.

We might come to the conclusion, then, that there

is not much to say of the universe after the $10^{27}$-year mark is reached. It will consist merely of supergalactic black holes in endless recession from each other (assuming, as most astronomers now think, that we live in an "open universe"; one, that is, that will expand forever) with a scattering of smaller black holes wandering through intercluster space. And, it might seem to us, there will be no significant change other than that expansion.

If so, we would probably be wrong.

The original feeling about black holes was that they were an absolute dead end—everything in, nothing out.

It seems, however, that is not so. The English physicist Stephen William Hawking (1942–    ), applying quantum-mechanical considerations to black holes, showed that they could evaporate. Every black hole has the equivalent of a temperature. The smaller the mass, the higher the temperature and the faster they will evaporate.

In fact, the rate of evaporation is inversely proportional to the cube of the mass, so that if black hole A is ten times as massive as black hole B, then black hole A will take a thousand times as long to evaporate. Again, as a black hole evaporates and loses mass, it evaporates faster and faster, and when it gets to be small enough, it evaporates explosively.

The temperature of sizable black hole is within a billionth of a billionth of a degree above absolute zero, so that their evaporation is dreadfully slow. Even after $10^{27}$ years, very little evaporation has yet taken place.

Indeed, what evaporation does take place is over-whelmed by the absorption of matter by the black holes as they swing through space. Eventually, though, little will remain to be absorbed and evaporation will slowly begin to dominate.

Very slowly, over eons and eons of time, the black holes shrink in size, the smaller ones shrinking faster. Then, one by one, in inverse order of size, they shrivel and pop explosively into oblivion. The really large black holes take $10^{100}$, or even $10^{110}$ years to do this.


In evaporating, black holes produce electromagnetic radiation (photons) and neutrino/antineutrino pairs. These possess no rest-mass, but only energy (which, of course, is a form of thinly spread-out mass).

Even if particles remain in space, they will not necessarily be permanent.

The mass of the universe is made up almost entirely of protons and neutrons, with a minor contribution from electrons. Until recently, protons (which make up about 95 percent of the mass of the universe right now) were thought to be completely stable, as long as they were left to themselves.

Not so, according to current theory. Apparently, protons can, very slowly, decay spontaneously into positrons, photons, and neutrinos. The half-life of a proton is something like $10^{31}$ years, which is an enormous interval, but not enormous enough. By the time all the black holes have evaporated, a so much longer time has elapsed that something like 90 percent of all the protons existing in the universe will have broken down. By the time $10^{32}$ years have passed, more than

99 percent of the protons will have broken down, and perhaps the black holes will also be gone by way of proton annihilation.

The neutrons, which can exist stably in association with protons, are liberated when protons break down. They are then unstable and, in the space of minutes, break down to electrons and protons. The protons then break down in their turn to positrons and massless particles.

The only particles remaining in quantity will then be electrons and positrons and in time they will collide, annihilating each other in a shower of photons.

By the time, then, that $10^{100}$ years have passed, the black holes will be gone, one way or the other. The universe will be a vast ball of photons, neutrinos, and antineutrinos, and nothing more, expanding outward indefinately. Everything will spread out more and more thinly, so that space will more and more approximate a vacuum.

One current theory, the so-called inflationary universe, begins with a total vacuum, one that contains not only no matter, but no radiation. Such a vacuum, according to quantum theory, can undergo random fluctuations to produce matter and antimatter in equal or nearly equal proportions. Generally, such matter/antimatter annihilates itself almost at once. Given time enough, however, a fluctuation may take place that will produce an enormously massive quanitity of matter/antimatter, with just enough unbalance to create a universe of matter in a sea of radiation. A superrapid expansion will then prevent annihilation and will produce a universe large enough to accommodate galaxies.

Perhaps, then, by the time, say, $10^{500}$ years pass, the universe will be close enough to a vacuum to allow fluctuations on a large scale to become possible again.

Then, amid the dead ashes of an old, old universe, a totally new one may be conceived, rush outward, form galaxies, and begin another long adventure. In that view (which I must admit I've made up myself and which has not been advanced by any reputable astronomer that I know of), the forever expanding universe is not necessarily a "one-shot" universe.

It may be that outside our universe (if we could reach outside to observe) there are the dregs of an enormously tenuous, enormously older universe, that faintly encloses us; and outside that a still more tenuous, far, far older one, that encloses both; and beyond that—forever and forever without end.


But what if we live in a "closed universe," one with a high enough density of matter in it to supply the gravitational pull required to bring the expansion to an end some day, and to begin a contraction, a falling together, of the universe?

The general astronomic view is that the density of matter in the universe in only about a hundredth of the minimum quantity required to close the universe, but what if the astronomers are wrong? What if the overall density of matter in the universe is actually twice the critical value?

In that case, it is estimated that the universe will expand till it is 60 eons old (four times its present age), at which time the slowing rate of expansion will have finally come to a halt. At that time, the universe

301

will have reached a maximum diameter of about 40 billion light-years.

The universe will then slowly start to contract and do so faster and faster. After another 60 eons, it will pinch itself into a Big Crunch, and finally disappear into the vacuum from which it originated.

Then, after a timeless interval, another such universe will form out of the vacuum—expand—and contract—over and over again without end. Or perhaps universes are formed in succession, some of which are open and some closed in random order.

No matter how we slice it, however, if we look far enough, we can end up with a vision of universe after universe, in infinite numbers through eternity—far as human eye could see.